Fisheries Science Series

This series delivers cutting-edge studies and accumulated wisdoms from Japan and Asia to the world. The aim of this series is to present new perspectives in fisheries science for the future of human welfare. As a country with a long history of fish-eating culture, Japan has created unique world-class cultures and technologies in fisheries, aquaculture, aquatic environment, seafood science, and other fishery-related sciences.This is an official book series of the Japanese Society of Fisheries Science.

More information about this series at http://www.springer.com/series/13529

Atsushi Hagiwara • Tatsuki Yoshinaga
Editors

Rotifers

Aquaculture, Ecology, Gerontology, and Ecotoxicology

Springer

JSFS

Editors
Atsushi Hagiwara
Graduate School of Fisheries and Environmental Sciences
Nagasaki University
Nagasaki, Japan

Tatsuki Yoshinaga
School of Marine Biosciences
Kitasato University
Sagamihara, Kanagawa, Japan

ISSN 2522-0470 ISSN 2522-0489 (electronic)
Fisheries Science Series
ISBN 978-981-10-5633-8 ISBN 978-981-10-5635-2 (eBook)
DOI 10.1007/978-981-10-5635-2

Library of Congress Control Number: 2017952235

Printed on acid-free paper

This Springer imprint is published by Springer Nature
The registered company is Springer Nature Singapore Pte Ltd.
The registered company address is: 152 Beach Road, #21-01/04 Gateway East, Singapore 189721, Singapore

Foreword to the Series

We all have to survive, and most of our food originates from that grown on land, but we can't overlook food from the sea. We catch creatures living in the water ecosystem by fishing techniques and eat them raw or cooked. That whole process and related activities are collectively called "fishery," and fishery is supported by fishery science that relates to a vast range of fields.

Fishery science brings us much knowledge—biological knowledge of the life in water; knowledge about their habitats and environment; knowledge to utilize these lives; political and administrative knowledge to organize social activities and system to distribute fishery products; technical and engineering knowledge of ships, fishing equipment, seaports, and harbors; and so on. It covers a great variety of subjects, and each subject contains both basic and applicative aspects relating to and essential to one another. To have fishery science prosper in human society, none of them can be ignored.

This series includes many of the aqua-bioscience fields and aquatic environment fields as the base of fishery science.

In this Fisheries Science Series, we provide you with carefully selected up-to-date topics of excellent works in the fields of fishery science. We hope our series can contribute to the development of fishery and the welfare of people worldwide.

Tokyo, Japan
July 2017

Katsumi Aida
Series Editor-in-Chief

Preface

Rotifers are microscopic metazoan ubiquitously found in aquatic environments, where they sustain the life of larger animals as food resource. Rotifers reproduce under a wide range of environments and show strong tolerance to environmental stress through their diapausing resting eggs. Through studies on their behavioral and physiological responses to varying environments, culture techniques of rotifers have substantially improved. This book aims to provide the most recent progress in rotifer studies in various fields in industry and academia. It is our hope that this book attracts interests of readers, including students and young researchers.

Among the 2300 rotifer species, this book mainly focused on the monogonont rotifer *Brachionus plicatilis*. Most rotifer species inhabit freshwater habitats, whereas *B. plicatilis* inhabits brackish and coastal marine waters and inland salt lakes. However, more than 70% of the scientific publications on rotifers have focused on this species. *B. plicatilis* is actually a species complex comprised of several ecologically distinct species. At the 14th International Rotifer Symposium in Ceske Budejovice, Czech Republic, in 2015, participating rotiferologists agreed on classifying *B. plicatilis* into 15 species (http://link.springer.com/article/10.1007/s10750-016-2725-7). The paper of the official description of species' names is now under preparation.

In Japan, the monogonont rotifer *B. plicatilis* was first known as a pest in eel culture ponds in the 1950s and 1960s. At that time, microalgae growth in eel culture ponds was promoted to reduce stress and enhance growth of eels. Rotifers are voracious grazers on microalgae, reproducing by cyclic parthenogenesis and eventually consuming most of the dissolved oxygen in the pond. This typically resulted in oxygen depletion in the eels, leading to their asphyxiation. Eel culturists could not find any effective means to control rotifers. Despite this unpromising beginning, rotifers are now recognized as a useful animal in industry and academia. In industry, rotifers have been utilized as initial live food for rearing many marine larval fish and crustaceans. Rotifers are also utilized in the wastewater treatment. Likewise, rotifers are used as model organisms in ecology, genetics, gerontology, and ecotoxicology.

We invited contributors of this volume from among the world's top scientists in this research. We also invited several young researchers from Japan. We anticipate that the readers of this book may be classified into two types: researchers and students in the area of aquaculture and basic science. Both fields are interlinked, and it is our hope that readers of this book can obtain comprehensive information about how rotifers are being employed in biological investigations. It is not unusual to find scientists from both fields studying similar topics using rotifers. In aquaculture, *Brachionus* has been an indispensable zooplankter since the 1960s, when Dr. Takashi Ito at Mie Prefectural University employed rotifers to feed fish larvae. Recognizing the importance of *Brachionus* in larviculture, their mass culture techniques have been intensively studied, and some essential achievements such as high-density culture, employment of valuable dietary algae, automated culture system, and effective production of resting eggs have been made. These have enabled stable and efficient aquatic seedling production for numerous important marine fish species. In addition, *Brachionus* is considered to be a suitable model organism for basic science research, because of its short life span, ease of culture, and its fascinating cyclical parthenogenetic life cycle. A series of studies with rotifers has significantly contributed to the understanding of life history evolution.

Basic information on rotifer biology is given in Part I, *Taxonomy and population genetics*. The genus *Brachionus* has been recognized since the 1700s, and its taxonomy and evolution have been controversial, especially before the employment of molecular markers. The current classification of the *B. plicatilis* species complex was described from two points of view: aquaculture and basic science. This part provides information for readers whose interests span from industry to academia.

Part II, *Live food*, provides essential information for readers who are interested in utilizing rotifers as live food in larviculture. *Brachionus* is the only reliable initial live food available for marine fish larvae, and its stable and efficient mass culture techniques have been mostly developed in Japan. The most recent techniques described in this part will be of great interest to aquaculturists.

Aside from its importance as live food, *Brachionus* has been employed as a model organism in various studies, from life history evolution to aging, ecotoxicology, and ecological diagnosis. These topics are included in Part III, *Model organism*, which provides recent progress in these areas. Recent developments in high-throughput DNA sequencing techniques have enabled us to obtain the whole genome sequence of *Brachionus*; thus, researches in rotifer genomics are now expected to further expanded.

The editors would like to thank all contributors of this book and reviewers who gave useful suggestions. Thanks are also extended to editors of *Fisheries Science Series*, Drs. Katsumi Aida, Hisashi Kurokura, Toyoji Kaneko, and Tadashi Tokai, the copy editor Yumi Terashima at the Japanese Society of Fisheries Science, Vignesh Iyyadurai Suresh, Chitra Sundarajan, Chieko Watanabe, and Mei Hann Lee at Springer Nature. We are also grateful to all scientists especially to the members of the "Rotifer Family" who participated in the past international rotifer symposia for providing us many insights in rotifer science. We also would like to express our greatest gratitude to researchers and technicians in fish hatcheries for providing us valuable information based on their own observations of their rotifer mass production systems. We would like to dedicate this volume to the rotifer species from all over the world for their exquisite beauty, the endless fascination, and giving us the opportunity to study them.

Nagasaki, Japan
Sagamihara, Japan
May 2017

Atsushi Hagiwara
Tatsuki Yoshinaga

Contents

Part II Live Food

Part I
Taxonomy and Population Genetics

Chapter 1
The Current Status of the Morphological Classification of Rotifer Strains Used in Aquaculture

Tomonari Kotani

Abstract Morphological variations of euryhaline rotifers *Brachionus*, including lorica size and form, have been recognized since they have been used as the live food for finfish larvae. Based on these morphological variations, rotifer populations were divided into different morphotypes. Finfish larvae ingest suitable size of feed fitting to their mouth size. Therefore, the size of rotifer fed to larval fishes should also be optimum for the larval mouth. Newly aquacultured fish species increased dramatically; thus, rotifers with optimum size for these species are required. In order to obtain such rotifers, the search for a new type or population and artificial manipulation to change the size has been attempted. Consequently, the newly found and developed rotifer species for aquaculture were applied to the larval rearing of fish species with small mouth. Up to date, cross-mating is the only method to develop new rotifer strains. Therefore, strict management is necessary to avoid the contamination among different rotifer species and strains.

1.1 Introduction

Euryhaline rotifer species belonging to genus *Brachionus* have been used as the feed for finfish or shellfish larvae since Ito (1960) succeeded in acclimatizing these rotifers to seawater and recommended their usage as marine finfish larval food. *Brachionus* species are known to show variation in size and are distinguished based on the shape or size of the lorica (Oogami 1976). Fish larvae have been found to prefer food sizes smaller than their mouth gap (Shirota 1970; Hagiwara 2008). Although rotifers have been known to have optimal size relative to the fish larvae mouth gap, small- or large-sized rotifers are occasionally found. Obtaining rotifers from the wild has been shown to be convenient to obtain small-sized rotifers for aquaculture. Hence, developing commercial rearing methods for the desired fish

T. Kotani (✉)
Faculty of Fisheries, Kagoshima University, Kagoshima, Japan
e-mail: kotani@fish.kagoshima-u.ac.jp

A. Hagiwara, T. Yoshinaga (eds.), *Rotifers*, Fisheries Science Series,
DOI 10.1007/978-981-10-5635-2_1

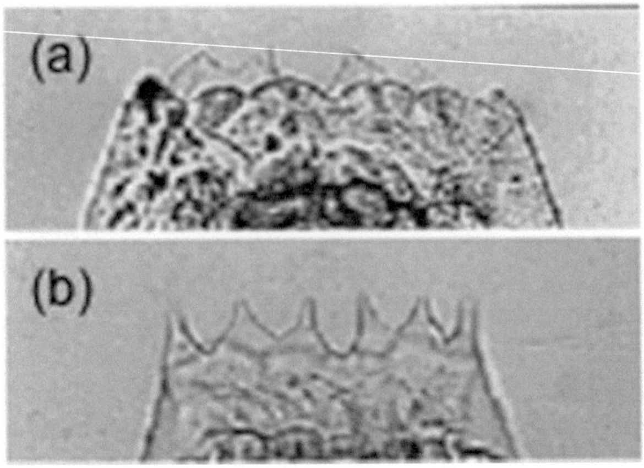

Fig. 1.1 Shape of the lorica spine of L- (**a**) and S-type (**b**) rotifers

species without obtaining appropriate-sized rotifers is difficult. In this chapter, the size variation and taxonomy of rotifers in relation to aquaculture and the methodology for applying rotifers of desired sizes are discussed.

1.2 Recognition of Size Variation

Brachionus plicatilis species complex had been classified as a single species (*Brachionus plicatilis*). They have a common morphological characteristic with three symmetrical pairs of lorica spine. *B. plicatilis* species complex can be distinguished from other *Brachionus* species on the basis of these spines. However, their populations isolated from various locations show large morphological variation. Rylov (1935) reported that *B. plicatilis* species complex, which is common in Europe, has two morphological types, i.e., longer type (Ferlängerte form) and wider type (Breite form). Hino and Hirano (1973) and Fukusho and Okauchi (1984) reported that large and small types of rotifers coexisted in the same culture ponds in the Japanese organizations where finfish larval rearing was conducted. Oogami (1976) classified this species complex based on the shape of a pair of central lorica spines and the size of lorica into two types: L type with larger lorica and obtuse-angled spine and S type with smaller lorica and acute-angled spine (Fig. 1.1). Fu et al. (1991a, b) collected 67 rotifer strains from around the world of which 37 and 30 were classified as S and L types, respectively, based on the shape of the spine. They investigated the morphometry of each type of strain and the genetic variation of the rotifer strains by using isozyme analysis. They confirmed that the classification by morphometry was consistent with the results of genetic analysis, and each classified group could be distinguished depending on the shape of the central lorica spines. By investigating the karyotype of three strains of L and S types each, Rumengan et al. (1991) found that L and S types had 22 and 25 chromosomes, respectively. Hirayama and Rumengan (1993) clarified that S type showed a different fecundity pattern from that of L type under similar culture conditions. Further, Hagiwara and Hirayama (1993) reported that the influence of temperature and salinity on sexual reproduction induction and fertilized egg formation was different between the two morphotypes. Moreover, Fu et al. (1993) reported the existence of reproductive isolation between them. Based on the findings of these studies, Segers (1995) concluded that *B. plicatilis* S and L types should be considered as different species. He suggested that S-type rotifers should be classified as *Brachionus rotundiformis* Tschugunoff and L-type ones should be referred to as *B. plicatilis* O. F. Müller.

On the other hand, since the 1980s, considerably smaller rotifers had been found from tropical areas, and they have been used for the larvae of fish species having a very small mouth size such as groupers, sillagos, and snappers (Lim 1993; Doi and Singhagraiwan 1993; Hagiwara et al. 1995). These rotifers have been referred to as SS type (Hagiwara et al. 1995). Hagiwara et al. (1995) found that SS-type rotifers were similar to *B. rotundiformis* in morphometric, ecological, and genetic characters, and there was no reproductive isolation between them. Kotani et al. (2005) analyzed the genetic differences among 67 rotifer strains, which had already been analyzed morphologically (Fu et al. 1991a) and genetically (Fu et al. 1991b), and identified seven more strains, including SS-type rotifers, by using isozyme and cluster analyses. The findings of these studies revealed that S- and L-type rotifers were divided into different population groups, whereas the SS type were included in S-type group. However, Kotani et al. (2005) also reported that the rotifer strains from tropical areas, including SS-type rotifers, formed one group within the S-type rotifer group (Fig. 1.2). Boehm et al. (2000) reported that SS-type rotifer clones could be discriminated with species-specific and strain-specific satellite DNA sequences. Hagiwara et al. (1995) found that the size and shape of lorica of SS-type rotifers were significantly different compared with those of a typical S type. Therefore, SS-type rotifers might be considered an independent species.

A study of taxonomic differences among rotifer clones found from temporary ponds in Spain (Gómez et al. 1995; Gómez and Snell 1996; Ciros-Pérez et al. 2001) classified into three types based on the morphometric and genetic analysis (Gómez et al. 1995). The clones were different in lorica size with the largest clone named as L, medium clone as SM, and the smallest as SS. Consequently, Ciros-Pérez et al. (2001) classified L and SS clones as *B. plicatilis* and *B. rotundiformis*, respectively. And they reclassified SM clones as neotype and described a new species named *B. ibericus*. Although *B. ibericus* could be significantly distinguished in lorica shape from *B. rotundiformis* and their lorica sizes had statistically significant difference, the variation of their lorica sizes overlapped. Similarly, *B. manjavacas* (Campillo et al. 2005; Fontaneto et al. 2007) could be distinguished morphometrically from other *Brachionus* species but their lorica size overlapped with that of *B. plicatilis* s.s. Hwang et al. (2013) reported that the morphometry of *B. koreanus* significantly differ from that of *B. plicatilis* and the lorica size of *B. koreanus* was also significantly smaller than that of *B. plicatilis*. Also *B. asplanchnoidis* was significantly different from *B. plicatilis* and *B. manjavacas* as the result of morphometry (Michaloudi et al. 2016). On the other hand, Gómez et al. (2002) suggested that the

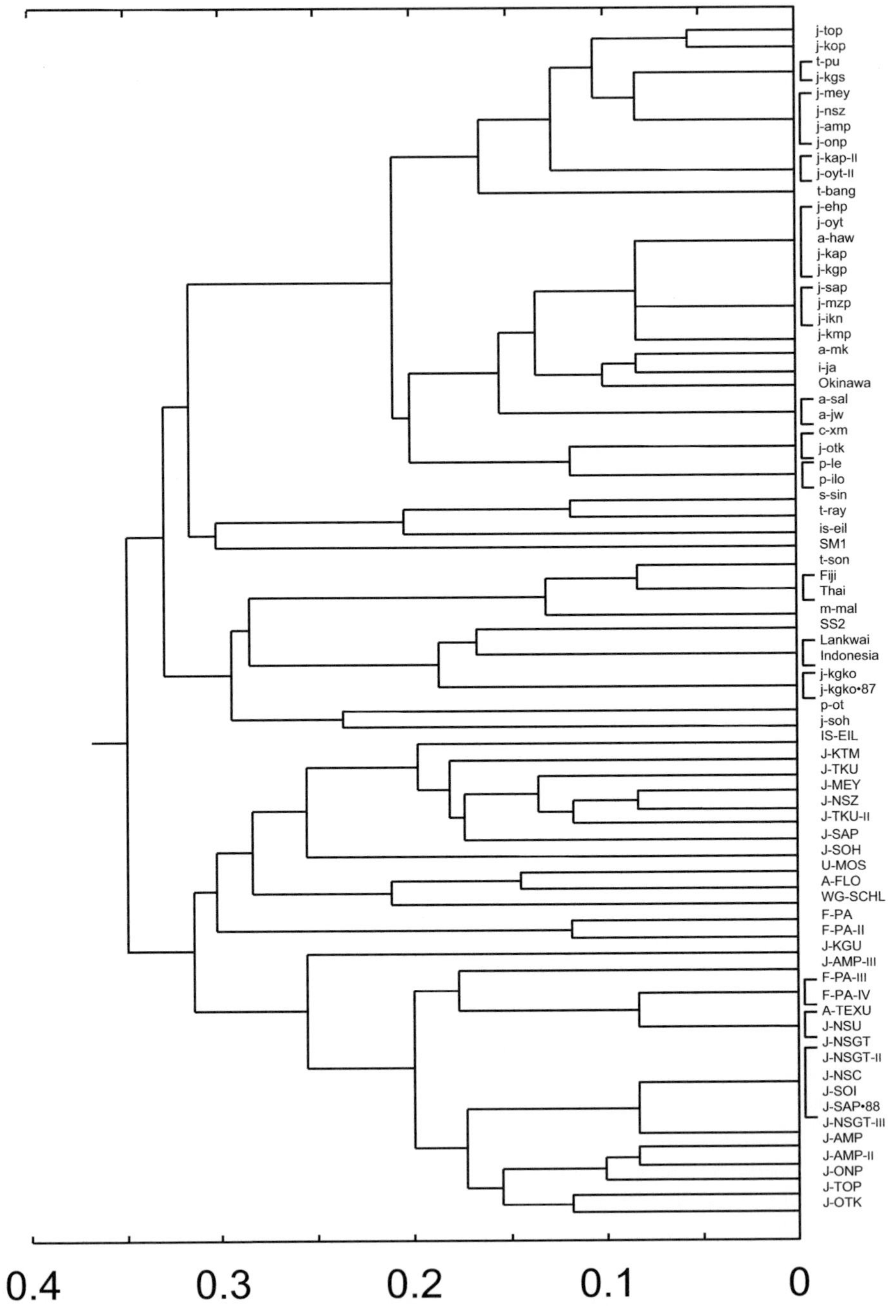

Fig. 1.2 Dendrogram for 74 strains according to a genetic distance index. The dendrogram is the result of the UPGMA analysis of Rogers' genetic distance (1972). *Lowercase* letters indicate the S morphotype; capital letters indicate the L morphotype (Kotani et al. 2005)

euryhaline rotifer *Brachionus* has at least nine phylogenic populations by conducting sequence analysis of two sites of the rotifer genome, COI and ITS1. Gómez et al. (2002) also introduced the unidentified *Brachionus* clones, Nevada, Cayman, Austria, Tiscar, and Almenara. Gómez and Snell (1996), Kostopoulou et al. (2009), and Malekzadeh-Viayeh et al. (2014) reported the morphological similarity or difference of these clones with known *Brachionus* species. Although these morphological analyses could indicate the difference among those *Brachionus* species or clones, the phylogeny among these *Brachionus* clones including *B. manjavacas*, *B. koreanus*, and *B. asplanchnoidis* was not independently determined with the morphological analysis but also was analyzed with DNA sequence analysis of COI and ITS1 (Campillo et al. 2005; Hwang et al. 2013; Malekzadeh-Viayeh et al. 2014; Michaloudi et al. 2016). Moreover, many studies reported that, even if DNA analysis was solely conducted, the classification of rotifer populations introduced previously could be valid (Papakostas et al. 2005, 2006a, b, 2009; Dooms et al. 2007; Gómez et al. 2007; Fontaneto et al. 2009; Vasileiadou et al. 2009). After all, based on these studies and the result of additional analysis, Mills et al. (2016) classified *Brachionus plicatilis* species complex into 15 species.

Brachionus Nevada and Cayman clones have already been used in some larval rearings (Papakostas et al. 2006b; Baer et al. 2008), and 15 new species of *Brachionus* rotifers are commonly used in hatcheries worldwide regardless of whether it is a known or new strain. As mentioned earlier, it is necessary to apply rotifers suitable for the mouth size of larvae during larval rearing. Therefore, aquaculturists require rotifers that are appropriate to the targeted fish species and hatcheries that have such rotifers. However, presently many hatcheries do not know the species that their rotifer strains belong to. In order to keep the rotifer strains with required characteristics in each hatchery, they have to know the classifications. Therefore, it is necessary to reclassify the strains found previously and used in hatcheries based on this classification. The reclassification of these strains might involve some difficulties, and it can be expected that the reclassification enables restricted management of rotifer strains.

1.3 Application for Fish Species Newly Developed in Aquaculture

New aquaculture fish species are developed every year, and the development of larval rearing methods for new species requires live food of appropriate size. SS-type rotifers, mentioned in the previous part, have been developed as the first live feed for groupers, sillagos, and snappers (Lim 1993; Doi and Singhagraiwan 1993; Hagiwara et al. 1995). Rotifers from tropical areas or Southern Europe are known to have the smallest size of all rotifer populations described so far. Their size is appropriate to small mouth sized larvae such as groupers and sillagos. Although S-type rotifers are used for some species of grouper larvae (Kawabe 1999; Teruya and Yoseda 2006), SS-type rotifers are recognized as important first live food for grouper larvae (Kayano and Wan 1997; Tanaka et al. 2005; Yoseda et al. 2006). Moreover, SS-type rotifers have been used for developing the larval rearing technique of new fish species for aquaculture, such as filefish *Stephanolepis cirrhifer*, blue-streak emperor *Lethrinus nebulosus*, clown fish *Amphiprion sebae*, and Japanese eel (Kimura et al. 2000; Kumar et al. 2010; Narita et al. 2011; Wullur et al. 2013). Since the introduction of SS-type rotifers as a finfish larval feed, more varieties of fish species can be cultured and increased the possibility to aquaculture other marine finfish species previously classified as unculturable.

Recently, larval rearing methods for red snapper, yellow tang, flame angelfish, and Napoleon fish were developed. However, the mouth sizes of these fishes are smaller than those of grouper larvae; therefore, SS-type rotifers could not be used for these fish larvae. Wullur et al. (2009) succeeded in rearing of seven-band grouper *Epinephelus septemfasciatus* by using minute monogonont rotifer *Proales similis*. The body length of this species is around 80 μm, and it is smaller than the SS-type rotifer. Hirai et al. (2012) confirmed the intake of *P. similis* by Napoleon fish larvae, which were reared using *P. similis* as live feed.

1.4 Determination of Appropriate Rotifer Strain for Finfish Larvae

In order for the fish larvae to grow properly, appropriate live food is necessary (Tanaka et al. 2005). Generally, if small-sized rotifers are not available, *Artemia* nauplii are used as feed. This is because finfish only consume feeds that are of appropriate size relative to their mouth size (Shirota 1970; Tanaka et al. 2005; Akazawa et al. 2008). Hence large-sized live food should be used to grow finfish juveniles. Akazawa et al. (2008) reported that larvae of *Verasper variegatus*, *Seriola quinqueradiata*, and *Platycephalus* sp. consumed larger rotifers, although the intensity of selectivity based on mouth size was different among these fish species. These three fishes required three types of rotifer as feed according to their average size. In most marine finfish hatcheries, not more than one type of rotifer is being used, using rotifer strains that have a size range that matches that of the reared larvae is necessary.

SS-type rotifers are fed to grouper larvae as the first feed. However, the supply of *Artemia* nauplii is not sufficient for the subsequent feeding after SS type of rotifer. The size gap between SS-type rotifers and *Artemia* nauplii is considerably large. Therefore, feeds with a size intermediate between SS-type rotifers and *Artemia* nauplii are necessary. Tanaka et al. (2005) reported that grouper larvae preferred large-sized feed for their growth. They fed L-type rotifers after SS-type rotifers and found improvement in the growth and survival of the larvae. Therefore, live feeds of adequate size should be used to rear finfish larvae. When excessively small live foods are fed, e.g., SS-type rotifer or *P. similis* mentioned above, intermediate-sized rotifers would be required subsequently. However, rotifer strains with the size between L-type rotifer and *Artemia* nauplii have not yet been found; this is a serious issue in the larviculture of finfish (Hagiwara 2008).

1.5 Artificial Modification of Rotifer Body Size

Larvae of groupers can be fed SS-type rotifers found in tropical area. However, larvae of red snapper and angelfish need to be fed considerably smaller live feeds than SS-type rotifers. Although protozoan and copepod nauplii were used as food candidates for these larvae, *P. similis* could be used as feed for fish larvae having a small mouth (Wullur et al. 2009). If a new species of finfish larvae with considerably smaller mouth are established for aquaculture, considerably smaller rotifers need to be produced. Further, for fish larvae with a larger mouth size, large-sized rotifers are required. Obtaining rotifer strains with specific characteristics, e.g., considerably smaller or larger size, requires artificial modification of body size, such as by breeding or chemical treatment.

When an animal with required characteristics is created artificially, the breeding protocols generally involve selection, mating, and gene manipulation. The euryhaline rotifer *Brachionus* is cyclical parthenogens; therefore, it can produce fertilized eggs in mictic generations. In the amictic generations, they undergo clonal reproduction. Therefore, the genetic diversity is low in one population, and selecting parental individuals within one population that show important characters becomes difficult. Fertilized eggs cannot be produced by rotifer species; however, they were found to be produced by S- and L-type rotifers (Fu et al. 1993; Kotani and Hagiwara 2003). Accordingly, mating among strains within species is used for breeding rotifers. Kotani et al. (2006) analyzed the lorica length, population growth rate, and induced mixis rate of the strain that hatched from fertilized eggs produced between two strains of L-type rotifers. The hybrid strain showed different characteristics from its parental strains. The size and growth rate of the hybrid strain were intermediate between the parental strains. Hence, if excessively large- or small-sized individuals are required, mating might not be an optimal method.

Hormonal treatment has also been attempted to vary the size of rotifers. Gallardo et al. (1999, 2000a, b) reported that hormonal treatment (e.g., growth hormone, juvenile hormone, and serotonin) could have some effects on reproduction and mixis induction of rotifers. Gallardo et al. (1997) found that treatment with gamma-aminobutyric acid and juvenile hormone could increase the lorica length of rotifers, resulting in an increase of size by 4–10%. No other study has used hormonal or chemical treatment for the enlargement or shortening of rotifer size. This might be because the effect of such treatment is temporary, and hatcheries require contemporary characteristics for rotifers. Hormonal or chemical treatments might be useful as temporary methods to obtain small-sized rotifers to be fed as the first food for larvae or to obtain large-sized rotifers for the duration until larvae are able to consume *Artemia* nauplii.

Therefore, obtaining rotifer strains with required sizes is possible via breeding or chemical treatment. However, when rotifers of smaller size such as that of SS type or the strain between L type and *Artemia* nauplius are required, breeding is not very useful since it can create strains with intermediate size between parental strains (Kotani et al. 2006), and chemical treatment can yield temporary effects (Gallardo et al. 1997). Further studies are warranted to overcome the shortcomings of breeding and chemical treatment for rotifers; at present, existing rotifer strains need to be used or new species that have the required size need to be identified from natural waters such as *P. similis*.

1.6 Prospects

One of the important characteristics of rotifers in larviculture of aquatic organisms is that they have optimum size for the mouth size of fish larvae. Since the mouth size of larvae differs across fish species, the required size of rotifers is also different across each species. If the mouth size of larvae is similar among fish species, one rotifer strain can be used for their larviculture. However, all rotifer strains cannot increase in size under conditions different from their original climate or environment. Hence, hatcheries have to obtain rotifer strain of appropriate fecundity and size for the targeted fish species. Moreover, based on molecular sequence divergence and concordant genetic patterns in cytochrome oxidase I and internal transcribed spacer 1, Mills et al. (2016) suggested that euryhaline rotifer *Brachionus* should be reclassified into 15 species. Since these are considered as different species, reproductive isolation among them is common. At present, mating is the only method to develop new rotifer strains. Further, accurately identifying rotifer species and cultivating them separately is necessary. If several strains need to be used to develop a new rotifer strain, strains that cannot hybridize with each other should not be selected. Even for strains that are not obtained by hybridization and were found in natural water, strict isolation needs to be confirmed, and hybridization among them should be prevented to maintain the original characteristics of each strain. Thus, strains with appropriate sizes for larvae should be selected, and isolation among them should be tightly controlled.

References

Akazawa A, Sakakura Y, Hagiwara A (2008) Feeding selectivity of marine fish larvae, *Verasper variegatus*, *Seriola quinqueradiata* and *Platycephalus* sp. on different sizes and shape of three rotifer strains. Nippon Suisan Gakkaishi 78:380–388. (in Japanese with English abstract)

Baer A, Langdon C, Mills S, Schulz C, Hamre K (2008) Particle size preference, gut filling and evacuation rates of the rotifer *Brachionus* "Cayman" using polystyrene latex beads. Aquaculture 282:75–82

Boehm EWA, Gibson O, Lubzens E (2000) Characterization of satellite DNA sequences from the commercially important marine rotifers *Brachionus rotundiformis* and *Brachionus plicatilis*. Mar Biotechnol 2:38–48

Campillo S, García-Roger EM, Martínez-Torres D, Serra M (2005) Morphological stasis of two species belonging to the L-morphotype in the *Brachionus plicatilis* species complex. Hydrobiologia 546:181–187

Ciros-Pérez J, Gómez A, Serra M (2001) On the taxonomy of three sympatric sibling species of the *Brachionus plicatilis* (Rotifera) complex from Spain, with the description of *B. ibericus* n. sp. J Plankton Res 23:1311–1328

Doi M, Singhagraiwan T (1993) Biology and culture of the red snapper, *Lutjanus argentimaculatus*. The research project of fishery resource development in the Kingdom of Thailand. Eastern Marine Fisheries Development Center (EMDEC), Rayong, p 51

Dooms S, Papakostas S, Hoffman S, Delbare D, Dierckens K, Triantafyllidis A, De Wolf T, Vadstein O, Abatzopoulos TJ, Sorgeloos P, Bossier P (2007) Denaturing gradient gel electrophoresis (DGGE) as a tool for the characterisation of *Brachionus* sp. strains. Aquaculture 262:29–40

Fontaneto D, Giordani I, Melone G, Serra M (2007) Disentangling the morphological stasis in two rotifer species of the *Brachionus plicatilis* species complex. Hydrobiologia 583:297–307

Fontaneto D, Kaya M, Herniou EA, Barraclough TG (2009) Extreme levels of hidden diversity in microscopic animals (Rotifera) revealed by DNA taxonomy. Mol Phylogenet Evol 53:182–189

Fu Y, Hirayama K, Natsukari Y (1991a) Morphological differences between two types of the rotifer *Brachionus plicatilis* O. F. Müller. J Exp Mar Biol Ecol 151:29–41

Fu Y, Hirayama K, Natsukari Y (1991b) Genetic divergence between S and L type strains of the rotifer *Brachionus plicatilis* O. F. Müller. J Exp Mar Biol Ecol 151:43–56

Fu Y, Hagiwara A, Hirayama K (1993) Crossing between seven strains of the rotifer *Brachionus plicatilis*. Nippon Suisan Gakkaishi 59:2009–2016

Fukusho K, Okauchi M (1984) Seasonal isolation between two strains of rotifer *Brachinous plicatilis* in an eel-culture pond. Nippon Suisan Gakkaishi 50:909–909

Gallardo WG, Hagiwara A, Tomita Y, Soyano K, Snell TW (1997) Effect of some vertebrate and invertebrate hormones on the population growth, mictic female production, and body size of the marine rotifer *Brachionus plicatilis* Müller. Hydrobiologia 358:113–120

Gallardo WG, Hagiwara A, Tomita Y, Snell TW (1999) Effect of growth hormone and gamma-aminobutyric acid on *Brachionus plicatilis* (Rotifera) reproduction at low food or high ammonia levels. J Exp Mar Biol Ecol 240:179–191

Gallardo WG, Hagiwara A, Snell TW (2000a) Effect of juvenile hormone and serotonin (5-HT) on mixis induction of the rotifer *Brachionus plicatilis* Muller. J Exp Mar Biol Ecol 252:97–107

Gallardo WG, Hagiwara A, Hara K, Soyano K, Snell TW (2000b) GABA, 5-HT and amino acids in the rotifers *Brachionus plicatilis* and *Brachionus rotundiformis*. Comp Biochem Phys A 127:301–307

Gómez A, Snell TW (1996) Sibling species and cryptic speciation in the *Brachionus plicatilis* species complex (Rotifera). J Evol Biol 9:953–964

Gómez Á, Temprano M, Serra M (1995) Ecological genetics of a cyclical parthenogen in temporary habitats. J Evol Biol 8:601–622

Gómez A, Serra M, Carvalho GR, Lunt DH (2002) Speciation in ancient cryptic species complexes: evidence from the molecular phylogeny of *Brachionus plicatilis* (Rotifera). Evolution 56:1431–1444

Gómez A, Montero-Pau J, Lunt DH, Serra M, Campillo S (2007) Persistent genetic signatures of colonization in *Brachionus manjavacas* rotifers in the Iberian Peninsula. Mol Ecol 16:3228–3240

Hagiwara A (2008) Zooplankton as a live feed for larval fish. In: Sugita H (ed) Diet and water for aquaculture. Koseisha Koseikaku, Tokyo, pp 59–99. (In Japanese)

Hagiwara A, Hirayama K (1993) Preservation of rotifers and its application in the finish hatchery. In: Lee CS, Su MS, Liao IC (eds) Finish hatchery in Asia: proceedings Finish Hatchery in Asia '91, TML Conf. Proc., vol 3, Tungkang Marine Lab., Taiwan Fisheries Research Inst, Tungkang, pp 66–71

Hagiwara A, Kotani T, Snell TW, Assava-Aree M, Hirayama K (1995) Morphology, reproduction, genetics, and mating behavior of small, tropical marine *Brachionus* strains (Rotifera). J Exp Mar Biol Ecol 194:25–37

Hino A, Hirano R (1973) On the size variation among *Brachionus plicatilis* populations. Abstracts of the 1973 spring meeting of the Japanese Society of Fisheries Science. p 73. In Japanese

Hirai N, Koiso M, Teruya K, Okuzawa K, Kobayashi M, Takebe T, Sato T, Nakamura K, Goto T, Hagiwara A (2012) Rearing conditions for Humphead Wrasse *Cheilinus undulatus* larvae, and introduction of the minute rotifer *Proales similis* as an initial live food. J Fish Sci Technol 4:57–64

Hirayama K, Rumengan IFM (1993) The fecundity patterns of S and L type rotifers of *Brachionus plicatilis*. Hydrobiologia 255/256:153–157

Hwang D-S, Dahms H-U, Park HG, Lee J-S (2013) A new intertidal *Brachionus* and intrageneric phylogenetic relationships among *Brachionus* as revealed by allometry and CO1-ITS1 gene analysis. Zool Stud 52:1–10

Ito T (1960) On the culture of mixohaline rotifer *Brachionus plicatilis* O. F. Müller in the sea water. Rep Fac Fish Prefectural Univ Mie 3:708–740. (in Japanese)

Kawabe K (1999) Evaluation of common S-type rotifer as an initial food for the larvae of blacktip grouper, *Epinephelus fasciatus*. Aquac Sci 47:403–408. (in Japanese with English abstract)

Kayano Y, Wan HY (1997) Initial feeding and growth of larval red spotted grouper *Epinephelus akaara*. Aquac Sci 45:213–218. (in Japanese with English abstract)

Kimura M, Motonaga F, Nakata Y, Nakamura S, Majikina M, Ishigaki A (2000) Production of larvae and juveniles of blue-streak emperor *Lethrinus nebulosus*. Ann Rep Okinawa Prefectural Sea Farming Cent 13:28–37. (in Japanese)

Kostopoulou V, Miliou H, Verriopoulos G (2009) Morphometric changes in a strain of the lineage 'Nevada', belonging to the *Brachionus plicatilis* (Rotifera) complex. Aquac Res 40:938–949

Kotani T, Hagiwara A (2003) Fertilization between the rotifer *Brachionus plicatilis* strains at different temperatures. Fish Sci 69:1078–1080

Kotani T, Hagiwara A, Snell TW, Serra M (2005) Euryhaline *Brachionus* strains (Rotifera) from tropical habitats: morphology and allozyme patterns. Hydrobiologia 546:161–167

Kotani T, Ihara K, Hagiwara A (2006) Cross-mating of euryhaline rotifer *Brachionus plicatilis* strains as a means to develop useful strains for larval food. Aquaculture 261:495–500

Kumar TTA, Setu SK, Murugesan P, Balasubramanian T (2010) Studies on captive breeding and larval rearing of clown fish [a[1]], *Amphiprion sebae* (Bleeker, 1853) using estuarine water. Indian J Mar Sci 39:114–119

Lim LC (1993) Larviculture of the greasy grouper *Epinephelus tauvina* F. and the brown-marbled grouper *E. fuscoguttatus* F. in Singapore. J World Aquac Soc 24:262–275

Malekzadeh-Viayeh R, Pak-Tarmani R, Rostamkhani N, Fontaneto D (2014) Diversity of the rotifer *Brachionus plicatilis* species complex (Rotifera: Monogononta) in Iran through integrative taxonomy. Zool J Linnean Soc 170:233–244

Michaloudi E, Mills S, Papakostas S, Stelzer C-P, Triantafyllidis A, Kappas I, Vasileiadou K, Proios K, Abatzopoulos TJ (2017) Morphological and taxonomic demarcation of *Brachionus asplanchnoidis* Charin within the *Brachionus plicatilis* cryptic species complex (Rotifera, Monogononta). Hydrobiologia 796:19–37

Mills S, Alcántara-Rodríguez AJ, Ciros-Pérez J, Gómez A, Hagiwara A, Galindo KH, Jersabek CD, Malekzadeh-Viayeh R, Leasi F, Lee J-S, Mark Welch DB, Papakostas S, Riss S, Segers H, Serra M, Shiel R, Smolak R, Snell TW, Stelzer C-P, Tang CQ, Wallace RL, Fontaneto D, Walsh EJ (2016) Fifteen species in one: deciphering the *Brachionus plicatilis* species complex (Rotifera, Monogononta) through DNA taxonomy. Hydrobiologia 796:39–58

Narita A, Kashiwagura M, Saito H, Okada Y, Akiyama N (2011) Effect of different rearing conditions on larval feeding activity, food consumption, survival and growth of filefish *Stephanolepis cirrhifer* larvae. Aquac Sci 59:551–561. (in Japanese with English abstract)

Oogami H (1976) On the morphology of *Brachionus plicatilis*. Newsl Form Izu Branch Shizuoka Prefectural Fish Res Cent 18:2–5. (in Japanese)

Papakostas S, Triantafyllidis A, Kappas I, Abatzopoulos TJ (2005) The utility of the 16S gene in investigating cryptic speciation within the *Brachionus plicatilis* species complex. Mar Biol 147:1129–1139

Papakostas S, Dooms S, Christodoulou M, Triantafyllidis A, Kappas I, Dierckens K, Bossier P, Sorgeloos P, Abatzopoulos TJ (2006a) Identification of cultured *Brachionus* rotifers based on RFLP and SSCP screening. Mar Biotechnol 8:547–559

Papakostas S, Dooms S, Triantafyllidis A, Deloof D, Kappas I, Dierckens K, Wolf TD, Bossier P, Vadstein O, Kui S, Sorgeloos P, Abatzopoulos TJ (2006b) Evaluation of DNA methodologies in identifying *Brachionus* species used in European hatcheries. Aquaculture 255:557–564

Papakostas S, Triantafyllidis A, Kappas I, Abatzopoulos TJ (2009) Clonal composition of *Brachionus plicatilis* s.s. and *B.* sp. 'Austria' hatchery strains based on microsatellite data. Aquaculture 296:15–20

Rogers JS (1972) Studies in genetics VII, University of Texas Publication, No. 7213. University of Texas, Austin, pp 145–153

Rumengan IFM, Kayano H, Hirayama K (1991) Karyotypes of S and L type rotifers *Brachionus plicatilis* O. F. Müller. J Exp Mar Biol Ecol 154:171–176

Rylov WM (1935) Das Zooplankton der Binnengewässer. In: Thienemann A (ed) Die Binnengewässer, vol XV. E. Schweizebart' sche Verlagsbuchhandlung, Stuttgart, pp 62–65

Segers H (1995) Nomenclature consequences of some recent studies on *Brachionus plicatilis* (Rotifera, Brachionidae). Hydrobiologia 313/314:121–122

Shirota A (1970) Studies on the mouth size of fish larvae. Nippon Suisan Gakkaishi 36:353–368. (in Japanese)

Tanaka Y, Sakakura Y, Chuda H, Hagiwara A, Yasumoto S (2005) Food selectivity of seven-band grouper *Epinephelus septemfasciatus* larvae fed different sizes of rotifers. Nippon Suisan Gakkaishi 71:911–916. (in Japanese with English abstract)

Teruya K, Yoseda K (2006) Successful mass production of early-stage larvae of kelp grouper *Epinephelus bruneus* in improved rearing conditions. Aquac Sci 54:187–194. (in Japanese with English abstract)

Vasileiadou K, Papakostas S, Triantafyllidis A, Kappas I, Abatzopoulos TJ (2009) A multiplex PCR method for rapid identification of *Brachionus* rotifers. Mar Biotechnol 11:53–61

Wullur S, Sakakura Y, Hagiwara A (2009) The minute monogonont rotifer *Proales similis* de Beauchamp: culture and feeding to small mouth marine fish larvae. Aquaculture 293:62–67

Wullur S, Yoshimatsu T, Tanaka H, Ohtani M, Sakakura Y, Kim H-J, Hagiwara A (2013) Ingestion by Japanese eel *Anguilla japonica* larvae on various minute zooplanktons. Aquac Sci 61:341–347

Yoseda K, Dan S, Sugaya T, Yokogi K, Tanaka M, Tawada S (2006) Effects of temperature and delayed initial feeding on the growth of Malabar grouper (*Epinephelus malabaricus*) larvae. Aquaculture 256:192–200

Chapter 2
Speciation in the *Brachionus plicatilis* Species Complex

Manuel Serra and Diego Fontaneto

Abstract The *Brachionus plicatilis* species complex is the best-studied example among rotifers where the use of integrative taxonomy, coupling morphology, ecology, physiology, cross-mating experiments, and DNA taxonomy helps disentangling the biological reality of the different species in the complex. Here we review the theoretical and empirical approaches in species definition applied to the *B. plicatilis* complex, we outline the history of the discovery of the complex, and we explore the evidence in support of the currently accepted presence of 15 species. We review the evidence for long-distance dispersal and for co-occurrence of the species in the complex, including the processes favoring the co-occurrence (e.g., niche differentiation). We discuss evolutionary explanations for the morphological similarity of the species in the complex. Finally, by reviewing studies on intraspecific population differentiation and mate recognition, we identify putative factors acting on speciation in the complex.

2.1 Introduction: Species and Speciation – An Overview with Stress on Rotifers

Consensus among biologists does not exist when they consider explicitly the species concept (e.g., Hey 2001; Pigliucci 2003; Wiens 2004a). Tens of species concepts have been developed and discussed until now since the definition of the Biological Species Concept by Mayr in 1942 (see reviews in de Queiroz 2005, 2007). Contrastingly, theoretical controversies based on different concepts are frequently absent when biologists deal with the concrete problem on what individuals should be identified as belonging to the same species or split in different ones (Hey 2006).

M. Serra (✉)
Cavanilles Institute of Biodiversity and Evolutionary Biology, University of Valencia, Valencia, Spain
e-mail: Manuel.Serra@uv.es

D. Fontaneto
Institute of Ecosystem Study, National Research Council, Verbania Pallanza, Italy
e-mail: d.fontaneto@ise.cnr.it

A. Hagiwara, T. Yoshinaga (eds.), *Rotifers*, Fisheries Science Series,
DOI 10.1007/978-981-10-5635-2_2

This suggests that an underlying, implicit consensus exists on the features the individuals of the same species have to share.

Regardless of species concepts, our knowledge on the speciation processes continuously improves (Coyne and Orr 2004; Nosil 2012) maybe even favored by the various and complementary approaches resulting from differences on how biologists conceive the species problem. The role of isolation (reproductive, geographical, ecological, temporal, etc.) is not disputed as the initial step of population differentiation (Coyne and Orr 2004). Such isolation represents a barrier to demographic exchange and gene flow between populations, and it then maintains the groups separated. Geographical isolation has been considered the most likely scenario for speciation, even if recent studies challenged the generality of the process, discussing the possibility of speciation with gene flow and in sympatry (Bolnick and Fitzpatrick 2007; Nosil 2008).

Small organisms impose a difficulty to the hypothesis of geographical isolation as one of the main mechanisms for speciation. The classical view was that organisms smaller than approximately 1 mm are ubiquitous, without biogeographic signatures (Finlay 2002; Fontaneto 2011). This would imply a difficulty for speciation, which would be consistent with the number of terrestrial animal species peaking around 1 mm (May 1978). Rotifers are microscopic, highly dispersible organisms due to the presence of long-surviving resting stages that can act as efficient propagules for dispersal (Artois et al. 2011). Consequently, it has been suggested that no effective geographical isolation may arise. In addition to high dispersal hampering isolation and population differentiation, the potentially high effective population size may be another hindrance to differentiation and speciation (Rossberg et al. 2013).

Notwithstanding the theoretical problems in species concepts and in the potential difference between larger and microscopic organisms, independently evolving entities are a reality of life, and even microscopic asexual bdelloid rotifers have been demonstrated to have "species" with patterns of morphological, ecological, genetic, and geographical differentiation similar to those in larger sexual organisms (Birky et al. 2005; Fontaneto et al. 2007b, 2008). Thus, there is ample support for being confident in using the term "species" for rotifers, with all the ecological and evolutionary implications of the term. It can be thought that species exist as discrete units of phenotypes because living beings face rather discrete environments (i.e., potential niches).

As suggested above, different species concepts seem to be loosely related to the practical criteria for detecting species and should not be confused with our theoretical understanding of the way species exist (Hey 2006). A criterion shared by sexual and asexual organisms, and thus by rotifers, is the fact that species have a genealogical and evolutionary ancestry, with coalescence of alleles in common ancestors (Avise and Ball 1990; Mallet 1995; Birky 2013). However, the common evolutionary criteria in our pragmatic approach should be that species are evolutionary independent entities, maintained at the same time separated between them and homogeneous within them by some cohesive force (Templeton 1989). Such cohesive force could be reproductive isolation for sexual organisms, but given that all bdelloid rotifers and several monogonont rotifers are strictly parthenogenetic, the idea of gene flow and reproductive isolation cannot hold true for them (Barraclough et al. 2009). Thus, the cohesive force should be something else, potentially ecological specialization and natural selection (Tang et al. 2014; Fontaneto and Barraclough 2015), commonly identified as the main drivers of ecological sweeps maintaining cohesion within species and differences between species in bacteria (Cohan 2002).

Cohesive processes create independent groups, i.e., species that can then be discovered by biologists using several approaches, including the traditional morphological approach in taxonomy, but also by universal sequence-based approaches, within the field of DNA taxonomy. The latter sequence-based approaches are becoming more and more important, as they provide not only data for the analysis of diversity patterns but also useful information to infer the evolutionary mechanisms behind the patterns. Species can thus be viewed as suitable tags supported by and useful to classify biological information, by means of complementary evidences from morphology, ecology, ethology, physiology, molecular biology, and genetic analysis in what can be called integrative taxonomy. Such continuous interaction between disciplines providing evidence of differences between groups is obvious in the case of the *Brachionus plicatilis* species complex. Before the taxonomical clarification of the group, there was an apparent confusion in the effects of environmental factors on what was called "*B. plicatilis*" tested in different studies, and this taxon was considered an extremely euryoecious species (Walker 1981). The reason was that different species were used in different studies. The taxon was misleadingly observed in a wide range of ecological conditions, and the strains provided differential responses to similar experimental setups in the laboratory.

Evidence of unexpected levels of species complex hiding more diversity than accepted accumulated in the last decade (Hebert et al. 2003; Tautz et al. 2003). As much of this evidence comes from molecular markers and phylogenies, several methods in DNA taxonomy have been developed to help in identifying cryptic species (e.g., Flot et al. 2010; Puillandre et al. 2012; Fujisawa and Barraclough 2013; Zhang et al. 2013). Cryptic species have been found in almost every group of animals (Knowlton 2000; Tang et al. 2012), and they seem to be homogeneously distributed among taxa and biogeographical regions (Pfenninger and Schwenk 2007). Much work has now been performed on genetic diversification in rotifers, and currently there are at least 42 known species complexes in rotifers (Fontaneto 2014; Gabaldón et al. 2017). We hypothesize that many more cases will be found. Surprisingly, almost every morphological rotifer species on which different tools of DNA taxonomy have been applied revealed cryptic diversity. The only morphological species that have been tested with a sufficient number of populations and for which no hidden diversity was found are *Rotaria socialis* among the bdelloids (Fontaneto et al. 2009) and *Keratella americana* among the monogononts (Hendrik Segers, personal communication). Thus, rotifers might be an exception to the homogeneous distribution of cryptic species in animal taxa, as they seem to harbor one of the highest levels of hidden diversity among animals, with complexes of tens of cryptic species (Fontaneto et al. 2009). Even genetic distances between species identified from DNA taxonomy in rotifers seems to be very high: genetic differences in rotifer species complexes in the barcoding locus COI are usually well above 5% between cryptic taxa and up to about 20% (Fontaneto 2014); these are much higher values than the commonly accepted 3% species threshold for COI in most animals (Hebert et al. 2003).

Therefore, the integration of morphological and DNA taxonomy, cross-mating experiments, and ecological and physiological evaluations in rotifers produced strong support for the separation of species and the interpretation of their ecological requirements. The best-known example of such detailed and fruitful analyses in rotifers is the case of the *Brachionus plicatilis* species complex.

2.2 *Brachionus plicatilis*: From a Species with Morphological Variability to a Species Complex

Within rotifers, *Brachionus plicatilis* Müller, 1786 (Monogononta, Brachionidae) is a textbook example of unclear morphological diversity, where only the use of integrative taxonomy, coupling morphology, ecology, physiology, cross-mating experiments, and DNA taxonomy could help disentangling the biological reality of the different species in the complex. *B. plicatilis* is a cyclical parthenogen whose populations proliferate via asexual reproduction with bouts of sexual reproduction resulting in the production of diapausing eggs. *B. plicatilis* is common in salt waters, easily grown in the laboratory, and used in aquaculture as larvae food; it attracted the interest of rotiferologists and achieved the status of a model organism (Yoshimura et al. 1996; Hagiwara et al. 2007; Stelzer et al. 2011; Kostopoulou et al. 2012).

Based on morphology, several names have appeared in the past, and some have been formally used as synonyms (e.g., *B. muelleri*, *B. hepatotomus*) or subspecies (*B. plicatilis asplanchnoidis*) (Koste 1978; Segers 1995; Jersabek et al. 2011). Therefore, classical taxonomists recognized this taxon as one with significant morphological variation (Koste 1978), and attempts to systematize it had been performed (Sudzuki 1987). As the traditional taxonomical work was based on preserved specimens collected in the field, this variability could not be split in their environmental and genetic components, although evidence for both plasticity and genetic effects started to be accumulated in the 1980s (Snell and Carrillo 1984; Serra and Miracle 1983). Oogami (1976), Snell and Carrillo (1984), and Fu et al. (1991a) began clarifying such variability, identifying two morphotypes (L, from large, and S, from small). The separation of these two morphotypes into distinct biological entities was supported by allozyme patterns (Fu et al. 1991b), chromosome number (Rumengan et al. 1991), the absence of mating between individuals of the two groups (Fu et al. 1993; Gómez and Serra 1995; Rico-Martinez and Snell 1995), and different mate recognition pheromones (Rico-Martinez and Snell 1995). Gómez et al. (1995) observed the co-occurrence in a small pond of three clusters of genotypes with different body size. These clusters did not show hybrids for their allozyme loci. Lack of hybridization in the field was interpreted as strong evidence for the existence of three species within the taxa called *B. plicatilis* at that time. Reproductive isolation between clusters was confirmed by mating experiments (Gómez and Serra 1995), and ecological differentiation was documented (Gómez et al. 1997).

Segers (1995) summarized all this in the initial work and identified the two morphologies with the formal nomenclatorial reestablishment of the names *Brachionus rotundiformis* Tschugunoff 1921 and *B. plicatilis* sensu stricto (s.s.) for the S and L morphotypes, respectively, within the *B. plicatilis* species complex. After Gómez et al. (1995) and additional morphological work (e.g., Yúfera 2001; Hagiwara et al. 1995), it became clear that more than two species existed in the complex. A first extensive view was provided in Gómez et al. (2002b). These authors, using mitochondrial (COI) and nuclear (ITS1) DNA, made phylogenetic inference on 50 specimens collected in 28 places. Sampling was concentrated in Spain (22 ponds). Despite this geographic restriction, nine clear-cut cluster with distances in COI equal or higher than 3% were detected. Thus, nine species were hypothesized. Interestingly, they are grouped in three major clades with differences in body size, and comparison with the congeneric species *B. calyciflorus* and *B. quadridentatus* suggested that the *B. plicatilis* species complex could be monophyletic. Suatoni et al. (2006) performed new phylogenetic and reproductive isolation analysis on a number of strains sampled around the world. Their results were in agreement with those in Gómez et al. (2002b), but as a result of the wider geographical sampling, they concluded that "possible up to 14 or more" species, with a minimum of seven, belong to the species complex. Additional phylogenetic work based on the same and/or different DNA sequences confirmed the occurrence of a high number of species in the complex (Baer et al. 2008; Gribble and Mark Welch 2012; Malekzadeh Viayeh et al. 2014) (Fig. 2.1). Mills et al. (2017), using the existent information in DNA databases and by obtaining new data, produced the most extensive phylogenetic analysis up to date. After applying DNA taxonomy, these authors proposed that 15 species are harbored in the complex. Currently, besides *B. plicatilis* s.s. and *B. rotundiformis*, the described species are *Brachionus ibericus* Ciros-Pérez et al. 2001b and *Brachionus koreanus* Wang et al. 2013, introducing two new, medium-sized morphotypes, *Brachionus manjavacas* Fontaneto et al. 2007 and *Brachionus asplanchnoidis* Charin 1947 as two additional species in the L-morphotype category (Ciros-Pérez et al. 2001b; Fontaneto et al. 2007a; Hwang et al. 2013; Michaloudi et al. 2017; Mills et al. 2017).

As the species in the complex are used in aquaculture as food for crustacean and fish larvae (e.g., Lubzens et al. 2001), a proper description of the species is most relevant for applied purposes. After recognition that aquaculturists were dealing with different species, methods for rapid molecular species identification have been developed (Papakostas et al. 2006), changes in the population dynamics in hatchery rotifer cultures can be better understood by monitoring if a species substitution has occurred (Papakostas et al. 2007), and selection of species for specific culture conditions and purposes can be optimized.

The finding of this rotifer species complex opened a number of interesting biological questions: Are the species in the complex widely distributed around the

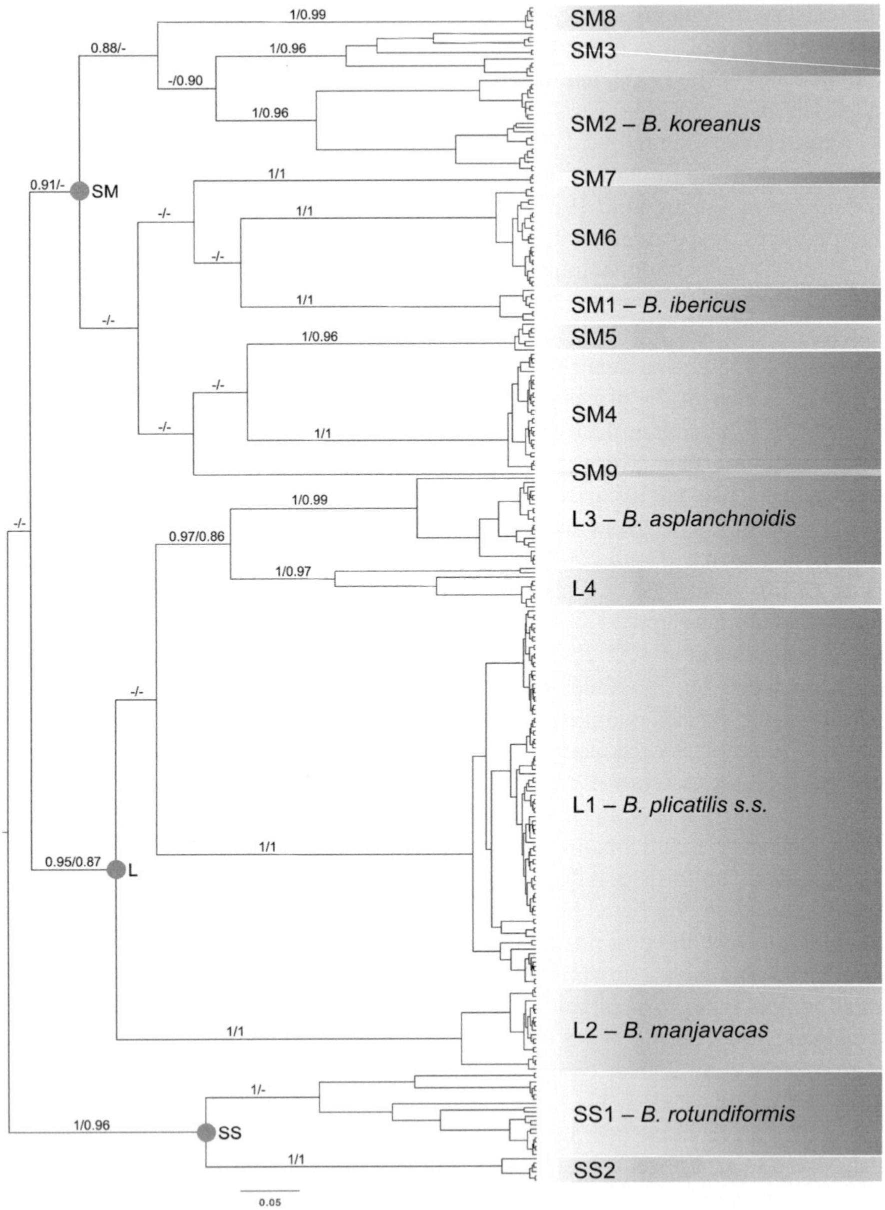

Fig. 2.1 Phylogenetic relationships in the *Brachionus plicatilis* species complex (Modified from Mills et al. 2017). A Bayesian inference reconstruction based on 275 COI haplotypes from 1223 individuals is shown. Posterior probabilities from BEAST/support values as approximate likelihood ratio test (aLRT) from PhyML are shown above each branch, but not for within-species branches; the '-' symbol indicates support <0.90 for posterior probabilities and <0.80 for aLRT tests. The three *gray circles* on basal nodes indicate the three main groups known in the species complex, namely, large (L), small-medium (SM), and small (SS). Clade names are numbered within the three main groups; species names for the six currently described species are also reported

world or restricted geographically? Are they ecologically specialized? Do they coexist in the same locations? If so, which are the processes allowing their co-occurrence? Do subtle morphological differences in the complex correlate with phylogeny? Is their morphological similarity due to recent speciation or to morphological stasis? Information more or less preliminary to answer to these questions already exists. On the other hand, the findings provide a frame to organize biological information, as that relevant in aquaculture, and ask for an appropriate specimen identification, based for instance on DNA barcoding, when reporting experiments or field observations requiring accurate taxonomy. Very relevantly, the findings are the first step to disentangle intraspecific and interspecific diversity and facilitate the study of population differentiation, population genetic structure, and the process of speciation.

2.3 Species Distribution and Co-occurrence

Even if more extensive sampling may be needed, current information suggests that long-distance dispersal of the cryptic species in the *Brachionus plicatilis* complex occurs (Gómez et al. 2002b; Mills et al. 2017). Within a single study or in different studies, isolates belonging to the same phylogenetic clade were obtained by sampling distant locations in the wild (Suatoni et al. 2006; Gómez et al. 2002b; Baer et al. 2008; Gribble and Mark Welch 2012; Malekzadeh Viayeh et al. 2014), and the same is suggested by the DNA database when combined with records associated to sampling (Mills et al. 2017). This suggests that differential species distribution would be more correlated with ecological factors (e.g., temperature, habitat type, or salinity) than with geographical and historical factors (i.e., species origin and barriers for dispersion). This contrasts with observations at the within-species level (see below). Additionally, several species can co-occur in the same region. For instance, it was early known that species of different morphotypes, now known to be different species, co-occur in Japan (Oogami 1976), and six species of the complex have been observed in the Iberian Peninsula (Gómez et al. 2002b).

Early observations on variation in body sizes in *Brachionus* populations (Serra and Miracle 1983; Fukusho and Okauchi 1984), reinterpreted after several species has been recognized, suggest now co-occurrence in a single locality with seasonal succession patterns. Using molecular methods co-occurrence in the same locality of several species of the complex was confirmed (Gómez et al. 1995; Papakostas et al. 2013). As an extreme case, Ortells et al. (2003) observed four species in a coastal lagoon, with the species having different but partially overlapping growing seasons. Cryptic species coexistence is not a rare phenomenon (Bickford et al. 2007) and poses a challenge to the competition theory of stable coexistence, since morphological similarity due to common evolutionary history is expected to correlate with niche similarity (Wiens 2004b; Losos 2008). If so, competitive exclusion should be a likely output. Investigation of the factors allowing co-occurrence of closely related species could shed light on when speciation would result in an increase of species diversity or contrarily in an exclusion of species with species turnover at evolutionary scale.

The species in the *B. plicatilis* complex tolerate a rather wide and similar range of salinity and temperature, but their population growth rates respond differentially within the tolerated range of these parameters (Gómez et al. 1997; Lowe et al. 2007; Montero-Pau et al. 2011; Gabaldón et al. 2013; Walczynska and Serra 2014). Broad tolerance is expected as a result of the adaptation to fluctuations of these parameters occurring in their habitats. However, small differences in the growth rate response to salinity may imply competitive exclusion depending on this condition (Lowe et al. 2007), so that the realized niche would be significantly lower than the fundamental niche (Soberón and Nakamura 2009). This implies a specialization, which makes possible alternative dominance periods of co-occurring species (e.g., seasonal substitution) in the water column. Interestingly, alternative dominance periods of species in the complex have been observed (Gómez et al. 1995; Ortells et al. 2003; Montero-Pau et al. 2011; Papakostas et al. 2013).

Species co-occurrence can be a transient phenomenon occurring before competitive exclusion arises or can be due to prolonged stable coexistence. These alternatives may occur at two different time scales: (a) within the growing season in the water column (i.e., during the parthenogenetic proliferation phase) or (b) involving the among-growing season dynamics (i.e., involving diapause). Focusing on the dynamics in the water column, Ciros-Pérez et al. (2001a) studied experimentally the competition dynamics of species in the complex differing in body size. They showed that pairs of species might coexist if food was diverse; otherwise exclusion occurred. Additionally, rotifer body size has implications for the vulnerability to predation by copepods (Lapesa et al. 2002, 2004), and differential vulnerability may compensate partially for advantages in competition (Ciros-Perez et al. 2004). Hence, with intrazooplanktonic predation, laboratory populations of these rotifer species co-occur longer than in the absence of such predation. Thus, differential use of resources and the trade-off between competitive capability and predation escape are potential factors mediating coexistence in the water column.

Rotifer populations are temporarily active in the ponds where they dwell, and their long-term persistence in a location depends on how each species matches its life cycle to its physical and biological environment. Rotifers like *Brachionus* have to produce diapausing eggs to survive adverse periods, which are recurrent in their habitats. The role of life cycle on competitor coexistence is stressed by the "storage effect," a mechanism that requires environmental fluctuations and is driven by the presence of a life stage that is protected against competition, e.g., the diapausing egg for *Brachionus* rotifers. The "storage effect" has been invoked to explain the co-occurrence of *B. plicatilis* and *B. manjavacas* (Montero-Pau et al. 2011), two cryptic species with very similar morphology (Campillo et al. 2005; Fontaneto et al. 2007a). Montero-Pau et al. (2011) suggested that fluctuations in salinity could be the environmental condition driving a stable coexistence. Additionally, recent theoretical research suggests that diversion of resources into sex and diapause weakens the otherwise superior competitor and could allow coexistence (Montero-Pau and Serra 2011). On the other hand, we know that some species in the complex differ in their pattern of sexual reproduction and in how the sexual, diapausing eggs leave diapause (Carmona et al. 1995; Walczynska and Serra 2014; Gabaldón et al. 2015). Escaping from competition via diapause and using diapause-leaving patterns to exploit time windows with low competition might offer persistence opportunities to the inferior competitors (Gabaldón et al. 2016).

Most of the work investigating stable coexistence of these species is theoretical, based on laboratory studies, or suggested by field patterns. Thus, more research is needed. For instance, it remains unknown if predation and differential resource exploitation are mechanisms involved in coexistence in the water column or if they rather promote temporary exclusion with species succession following fluctuation of these factors. Moreover, because some species have a wide niche overlap (Gabaldón et al. 2013) with similar growth rates, co-occurrence in the same pond due to a lasting, slow dynamics toward exclusion cannot be ruled out.

When two phylogenetically close species co-occur, differentiation in mating systems has important implications. In the absence of pre-zygotic isolation, if post-zygotic isolation is present, the less frequent species is driven to local extinction because it is allocating relatively more gametes in wrong mates – a "gamete sink" – while if post-zygotic isolation is absent, hybridization will cause the lineages to merge. Intermediate cases with limited genetic introgression between species are still possible. Pre-zygotic isolation between monogonont rotifer species needs differentiation in the timing of the sexual phases or accurate mate recognition of cospecific mates. Molecular evidence shows that introgression has occurred between species in the *Brachionus plicatilis* complex (Gribble and Mark Welch 2012), which is consistent with rare but existing laboratory observations of copulations between males and females of different species (Suatoni et al. 2006).

2.4 Morphological Evolution in the Complex

As far as we know, phylogenetic divergence correlates partially with the subtle morphological divergence occurring in the *B. plicatilis* complex. First, the complex seems to be monophyletic (Mills et al. 2017), which means that traditional, morphologically based taxonomy was able to correctly identify monophyly. Second, as stated above, the three major morphologies (large, intermediate, and small) are also three major clades in the phylogenetic analysis based on DNA (e.g., Gómez et al. 2002b; Mills et al. 2017). This suggests that the morphotypes are also monophyletic, and that the major size divergence in the complex evolved only once. We know now that body size correlates to genome size, which has high variation in the species complex (range about sevenfold; Stelzer et al. 2011; see also Mills et al. 2017).

The high levels of cryptic diversity in the *B. plicatilis* complex and in most of other rotifer complexes, together with the high genetic distances between cryptic species in the complex, much higher than in other animals, suggest two potential scenarios. On one hand, exceptional morphological stasis could be responsible for the observed pattern of discrepancy between morphology and DNA taxonomy; on

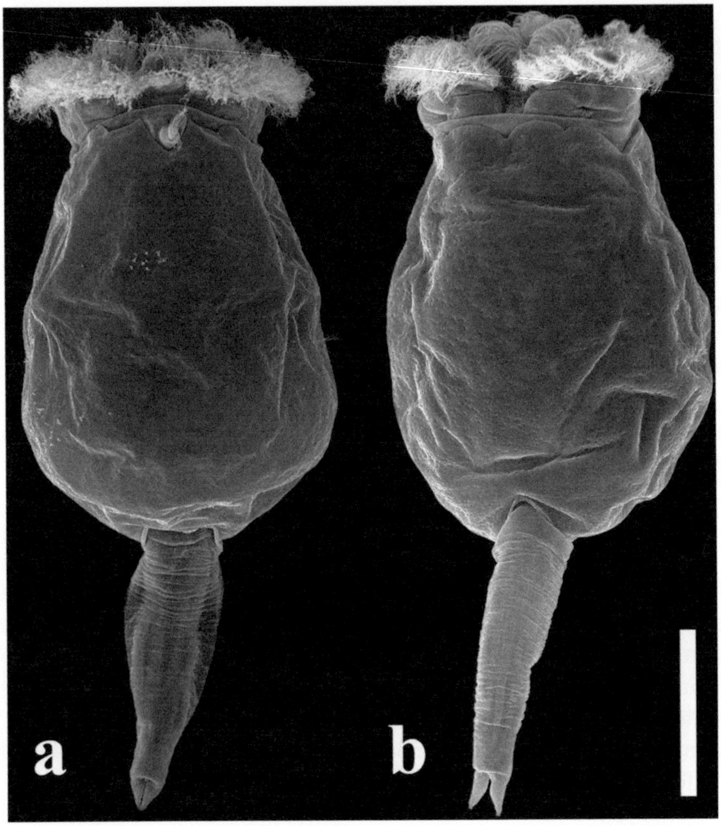

Fig. 2.2 *Brachionus manjavacas* in dorsal (**a**) and ventral (**b**) view. Scale bar = 50 μm (Modified from Fontaneto et al. 2007a)

the other hand, it is possible that rotifers harbor high speciation rates, allowing for separation of independently evolving entities in a relatively short time.

Nothing is really known in diversification rates in rotifers, except that the very approximate estimates of the speciation events within the *B. plicatilis* complex are in the range of 20–30 million years ago (Gómez et al. 2002b; Fontaneto et al. 2012). A critical case is that of *B. manjavacas* (Fig. 2.2) and *B. plicatilis* s.s., which are practically indistinguishable even by biometrical analyses (Campillo et al. 2005; Fontaneto et al. 2007a), but are estimated to have diverged more than 20 million years ago based on COI. The estimates on divergence times were obtained without any fossil calibration, impossible for rotifers and their sister phyla, due to lack of fossil evidence. Thus, estimates were calibrated using the general evolutionary rates in the analyzed loci. Yet, both mitochondrial and nuclear calibrations suggested similar ages (Gómez et al. 2002b), supporting their reliability. If such old ages are realistic, speciation events in the species complex are not recent at all, but date back as much as to the Oligocene epoch. Speciation events between closely related sister taxa in other animals are usually much more recent, for example, in birds and mammals, around three million years near the equator and one million years at higher latitudes (Weir and Schluter 2007) and in North American tiger beetles, less than two million years (Barraclough and Vogler 2002). Diversification events dating back to 20 million years usually date divergence between genera and subgenera in other animals (e.g., Hines 2008; Andujar et al. 2012).

Thus, morphological stasis – and not recent speciation events – seems to be a more plausible explanation for the origin of the species complex. Morphological stasis in complexes of cryptic species is not a rare phenomenon, but the evolutionary mechanism underlying it is not yet understood (Knowlton 2000; Bickford et al. 2007).

2.5 Population Differentiation and Speciation

A first step in the geographical model of speciation is genetic differentiation among populations of a single species (Futuyma 1998). Therefore, documenting population differentiation is important to assess if and how such allopatric speciation could occur in monogonont rotifers. Population differentiation can occur by neutral evolutionary processes as founder effects or genetic drift, by adaptation to local conditions, or by a combination of neutral and selective processes. In any case, effective gene flow opposes to population divergence.

Zooplankters inhabiting ponds and lakes have high dispersal capability (Jenkins and Buikema 1998). Thus, gene flow is expected to be high and population differentiation to be low. Contrasting with this expectation, molecular marker studies have shown that aquatic invertebrates have high within-species genetic differentiation. This disagreement is called the "dispersal-gene flow paradox" (De Meester et al. 2002). The species in the *B. plicatilis* complex fit in this paradox. Using mitochondrial DNA, Gómez et al. (2000, 2007) described the patterns of population differentiation in *B. plicatilis* s.s. and *B. manjavacas* in the Iberian Peninsula. Deep phylogenetic clades were found within each species. Phylogeographic analysis suggests that allopatric fragmentation during glaciations and post-glaciation geographic range expansion occurred. Both species have similar phylogeographic patterns in the Iberian Peninsula. Microsatellite studies on *B. plicatilis* populations in the Iberian Peninsula confirmed deep genetic divergence within this species (Gómez et al. 2002a, b). Derry et al. (2003) also found differentiation among Canadian populations of *B. plicatilis* inhabiting lakes located in an 85-km radius area. These results confirm low gene flow among populations in the same region.

To explain genetic differentiation in high-dispersal organisms, Boileau et al. (1992) suggested a role of founder effects, which would be persistent due to the large population size of these organisms. Large population size would cause a "high-density blocking" (Hewitt 1993), so that immigrants after population foundation would have low effect. A role of local adaptation in acting against new immigrants was later suggested as an additional factor in the so-called monopolization hypothesis (De Meester et al. 2002). Studies on local adaptation within species of the *B. plicatilis* complex are scarce. However, we know that the life history of *B. plicatilis* s.s. populations respond differently to the experimental temperature and salinity (Campillo et al. 2009, 2011). When the differential response is related to the local conditions of each population, it suggests cases of local adaptation to salinity and habitat temporality, particularly in the sexual reproduction patterns. Interestingly, Papakostas et al. (2013) found a within-species succession in Lake Koronia suggesting differential adaptation of genotypes to seasonal conditions. However, between-population divergence in the life-history traits is not correlated to between-population divergence in neutral markers (Campillo et al. 2009). This lack of an "isolation-by-adaptation" pattern (Nosil 2008) introduces a doubt about the role of local adaptation in avoiding gene flow in rotifers. Rotifers have much higher population sizes than other zooplankters, so "high-density blocking" might be the dominant factor in decreasing effective gene flow.

The results summarized here and those in Sect. 3 point out that *B. plicatilis* species have not important geographic barriers avoiding distant places to be colonized, but they have within-species phylogeographic patterns. These are not contradictory results. If pre-zygotic isolation does not occur, being the residents' and the immigrants' niches very similar, the immigrants would affect the local gene pool only if they have higher fitness than the residents, which is expected to be infrequent; otherwise, the "high-density blocking" or the "gamete sink" would act against immigrants. If pre-zygotic isolation between residents and immigrants occurs, immigration might be successful due to (1) a higher fitness of the immigrants causing exclusion of the residents or (2) niche partitioning.

In the *B. plicatilis* species complex, mate recognition is mediated by the so-called mate recognition pheromone (MRP), which is a glycoprotein located on the female body surface (Snell et al. 1995, 2009; Snell and Stelzer 2005). MRP is encoded by the *mrp-b* gene family (Snell et al. 2009; Gribble et al. 2011). Gribble and Mark Welch (2012) studied the relationship between *mrp-b* distance and the reproductive isolation measured in behavioral tests (e.g., copulation frequency). They used 11 phylotypes (likely species) in the *B. plicatilis* complex and found a positive correlation between divergence of *mrp-b* and reproductive isolation. Evidence for positive selection on *mrp-b* was not found. Their analysis suggests that *mrp-b* is evolving rapidly, and that this evolution could result in saltational speciation in a neutral way. This opens the possibility for sympatric speciation to occur in the complex. If such isolation could emerge, the new species might be stabilized in sympatry by seasonal specialization, a phenomenon observed in *B. plicatilis* population of Lake Koronia (Papakostas et al. 2013).

2.6 Prospects

This short review of the history of the discovery of the hidden diversity in the *B. plicatilis* complex, together with all the supporting evidence for the existence of several species, clearly shows that our current knowledge on the complex is mature to clarify most of the taxonomic ambiguities. It is true that morphology may not be completely reliable for a proper identification of species in the complex, but time has come to provide unambiguous names to be correctly used for all the taxa in the complex. A lack in clarity for the use of names will only create confusion, whereas an unambiguous and official name for all the taxa in the complex will greatly help communication and comparison of results between different studies. Six species have already been named, whereas others received only an unofficial name: there is a strong and urgent need to stabilize the nomenclature in the complex, and all the unofficial names should be made available and the species described according to the International Code of Zoological Nomenclature. Thus, all the unofficially named taxa and the ones that still have no name should receive a formal taxonomic description, based on all the evidences that have been gathered until now, combining morphology, DNA taxonomy, ecology, physiology, and cross-mating experiments. This will allow for a clarification of the nomenclature and taxonomy in the group and will pose the basis for avoiding confusion in the future.

Further than providing taxonomically and nomenclaturally valid names for all the taxa in the complex, a more detailed characterization of the ecology, behavior, and physiology of the different species will improve our understanding of the processes that may have acted in driving such high differentiation. Disentangling and inferring the processes will take more time than describing the patterns, but will provide highly valuable scientific insights into biological evolution. Moreover, given that *Brachionus* rotifers have fast demographic and reproductive rates, they could be used as models for evolutionary experiments in speciation. Until now, such studies have been performed only in prokaryotes, due to their obvious advantages in lab cultures; nevertheless, the *Brachionus plicatilis* species complex is a very suitable candidate to perform experiments on animals.

In order to infer the processes behind diversification events in the complex, also a more detailed knowledge on the geographic distribution of the different taxa will be helpful. Thus, given that sequencing tools and techniques become continuously quicker, more reliable, and cheaper, we envisage that a large amount of information will be gathered on common molecular markers for several additional populations in different parts of the world. The expectation is that more species would also be discovered in the complex, especially from remote and understudied habitats.

Overall, the impact of the ecological and evolutionary studies on the *B. plicatilis* complex triggered a large amount of similar studies on other rotifers. Currently, species complexes in other rotifers and in other *Brachionus* are under scrutiny, and they all seem to show a similar pattern to the one described in the *B. plicatilis* complex. Studies on *B. plicatilis* are at a more advanced step, given that more knowledge has already been accumulated compared to other species complexes. Thus, *B. plicatilis* as a model system will still be at the forefront of rotifer studies for a long time, and new methods and developments originating from it will then be applied and used also for other models, providing a clearer picture of the diversification processes in rotifers and in animals in general.

Acknowledgments We thank T.W. Snell, M.J. Carmona, and E. Gracía-Moisés for their comments on the MS. MS research was supported by a grant from the Spanish Ministry of Science (CGL2009- 07364).

References

Andujar A, Serrano J, Gomez-Zurita J (2012) Winding up the molecular clock in the genus *Carabus* (Coleoptera: Carabidae): assessment of methodological decisions on rate and node age estimation. BMC Evol Biol 12:40

Artois T, Fontaneto D, Hummon WD, McInnes SJ, Todaro MA, Sørensen MV, Zullini A (2011) Ubiquity of microscopic animals? Evidence from the morphological approach in species identification. In: Fontaneto D (ed) Biogeography of microscopic organisms, is everything small everywhere? Cambridge University Press, Cambridge, pp 244–283

Avise JC, Ball RM Jr (1990) Principles of genealogical concordance in species concepts and biological taxonomy. Oxf Surv Evol Biol 7:45–67

Baer A, Langdon C, Mills S, Schulz C, Hamre K (2008) Particle size preference, gut filling and evacuation rates of the rotifer *Brachionus* 'Cayman' using polystyrene latex beads. Aquaculture 282:75–82

Barraclough TG, Vogler AP (2002) Recent diversification rates in North American Tiger Beetles estimated from a dated mtDNA phylogenetic tree. Mol Biol Evol 19:1706–1716

Barraclough TG, Fontaneto D, Herniou EAm Ricci C (2009) The evolutionary nature of diversification in sexuals and asexuals. In: Butlin R, Bridle J, Schluter D (eds) Speciation and patterns of diversity. Cambridge University Press, Cambridge, pp 29–45

Bickford D, Lohman DJ, Sodhi NS (2007) Cryptic species as a window on diversity and conservation. Trends Ecol Evol 22:148–155

Birky CW Jr (2013) Species detection and identification in sexual organisms using population genetic theory and DNA sequences. PLoS One 8:e52544

Birky CW Jr, Wolf C, Maughan H, Herbertson L, Henry E (2005) Speciation and selection without sex. Hydrobiologia 546:29–45

Boileau MG, Hebert PDN, Schwartz SS (1992) Nonequilibrium gene frequency divergence – persistent founder effects in natural populations. J Evol Biol 5:25–39

Bolnick DI, Fitzpatrick BM (2007) Sympatric speciation: models and empirical evidence. Annu Rev Ecol Evol Syst 38:459–487

Campillo S, Garcia-Roger EM, Martinez-Torres D, Serra M (2005) Morphological stasis of two species belonging to the L-morphotype in the *Brachionus plicatilis* species complex. Hydrobiologia 546:181–187

Campillo S, García-Roger EM, Carmona MJ, Gómez A, Serra M (2009) Selection on life-history traits and genetic population divergence in rotifers. J Evol Biol 22:2542–2553

Campillo S, García-Roger EM, Carmona MJ, Serra M (2011) Local adaptation in rotifer populations. Evol Ecol 25:933–947

Carmona MJ, Gómez A, Serra M (1995) Mictic patterns of the rotifer *Brachionus plicatilis* Müller 1786 in small ponds. Hydrobiologia 313/314:365–371

Ciros-Pérez J, Carmona MJ, Serra M (2001a) Resource competition between sympatric sibling rotifer species. Limnol Oceanogr 46:1511–1523

Ciros-Pérez J, Gómez A, Serra M (2001b) On the taxonomy of three sympatric sibling species of the *Brachionus plicatilis* (Rotifera) complex from Spain, with the description of *B. ibericus* n.sp. J Plankton Res 23:1311–1328

Ciros-Perez J, Carmona MJ, Lapesa S, Serra M (2004) Predation as a factor mediating resource competition among rotifer sibling species. Limnol Oceanogr 49:40–50

Cohan FM (2002) What are bacterial species? Annu Rev Microbiol 56:457–487

Coyne JA, Orr HA (2004) Speciation. Sinauer Associates, Sunderland. 545 pp

De Meester L, Gómez A, Okamura B, Schwenk K (2002) The monopolization hypothesis and the dispersal-gene flow paradox in aquatic organisms. Acta Oecol 23:121–135

de Queiroz K (2005) Different species problems and their resolution. BioEssays 27:1263–1269

de Queiroz K (2007) Species concepts and species delimitation. Syst Biol 56:879–886

Derry AM, Hebert PDN, Prepas EE (2003) Evolution of rotifers in saline and subsaline lakes: a molecular phylogenetic approach. Limnol Oceanogr 48:675–685

Finlay BJ (2002) Global dispersal of free-living microbial eukaryote species. Science 296:1061–1063

Flot J-F, Couloux A, Tillier S (2010) Haplowebs as a graphical tool for delimiting species: a revival of Doyle's "field for recombination" approach and its application to the coral genus *Pocillopora* in Clipperton. BMC Evol Biol 10:372

Fontaneto D (ed) (2011) Biogeography of microscopic organisms, is everything small everywhere? Cambridge University Press, Cambridge

Fontaneto D (2014) Molecular phylogenies as a tool to understand diversity in rotifers. Int Rev Hydrobiol 99:178–187

Fontaneto D, Barraclough TG (2015) Do species exist in asexuals? Theory and evidence from bdelloid rotifers. Integr Comp Biol 55:253–263

Fontaneto D, Giordani I, Melone G, Serra M (2007a) Disentangling the morphological stasis in two rotifer species of the *Brachionus plicatilis* species complex. Hydrobiologia 583:297–307

Fontaneto D, Herniou EA, Boschetti C, Caprioli M, Melone G, Ricci C, Barraclough TG (2007b) Independently evolving species in asexual bdelloid rotifers. PLoS Biol 5:914–921

Fontaneto D, Barraclough TG, Chen K, Ricci C, Herniou EA (2008) Molecular evidence for broad-scale distributions in bdelloid rotifers: everything is not everywhere but most things are very widespread. Mol Ecol 17:3136–3146

Fontaneto D, Kaya M, Herniou EA, Barraclough TG (2009) Extreme levels of hidden diversity in microscopic animals (Rotifera) revealed by DNA taxonomy. Mol Phylogenet Evol 53:182–189

Fontaneto D, Tang CQ, Obertegger U, Leasi F, Barraclough TG (2012) Different diversification rates between sexual and asexual organisms. Evol Biol 39:262–270

Fu Y, Hirayama K, Natsukari Y (1991a) Morphological differences between two types of the rotifer *Brachionus plicatilis* O. F. Müller. J Exp Mar Biol Ecol 151:29–41

Fu Y, Hirayama K, Natsukari Y (1991b) Genetic divergence between S and L type strains of the rotifer *Brachionus plicatilis* O. F. Müller. J Exp Mar Biol Ecol 151:43–56

Fu Y, Hagiwara A, Hirayama K (1993) Crossing between seven strains of the rotifer *Brachionus plicatilis*. Nippon Suisan Gakkaishi 59:2009–2016

Fujisawa T, Barraclough TG (2013) Delimiting species using single-locus data and the Generalized Mixed Yule Coalescent (GMYC) approach: a revised method and evaluation on simulated datasets. Syst Biol 62:707–724

Fukusho K, Okauchi M (1984) Seasonal isolation between two strains of rotifer *Brachionus plicatilis* in an eel-culture pond. Bull Jpn Soc Sci Fish 50:909

Futuyma DJ (1998) Evolutionary biology, 3rd edn. Sinauer Associates, Sunderland

Gabaldón C, Montero-Pau J, Serra M, Carmona MJ (2013) Morphological similarity end ecological overlap in two rotifer species. PLoS One 8:Le57087

Gabaldón C, Serra M, Carmona MJ, Montero-Pau J (2015) Life history traits, abiotic environment and coexistence: the case of two cryptic rotifer species. J Exp Mar Biol Ecol 465:142–152

Gabaldón C, Fontaneto D, Carmona MJ, Montero-Pau J, Serra M (2016) Ecological differentiation in cryptic rotifer species: what we can learn from the *B. plicatilis* complex. Hydrobiologia. doi:10.1007/s10750-016-2723-9

Gómez A, Serra M (1995) Behavioural reproductive isolation among sympatric strains of *Brachionus plicatilis* Müller, 1786: insights into the status of this taxonomic species. Hydrobiologia 313/314:111–119

Gómez A, Temprano M, Serra M (1995) Ecological genetics of a cyclical parthenogen in temporary habitats. J Evol Biol 8:601–622

Gómez A, Carmona MJ, Serra M (1997) Ecological factors affecting gene flow in the *Brachionus plicatilis* complex (Rotifera). Oecologia 111:350–356

Gómez A, Carvalho GR, Lunt DH (2000) Phylogeography and regional endemism of a passively dispersing zooplankter: mitochondrial DNA variation in rotifer resting egg banks. Proc R Soc Lond B 267:2189–2197

Gómez A, Adcock GJ, Lunt DH, Carvalho GR (2002a) The interplay between colonization history and gene flow in passively dispersing zooplankton: microsatellite analysis of rotifer resting egg banks. J Evol Biol 15:158–171

Gómez A, Serra M, Carvalho GR, Lunt DH (2002b) Speciation in ancient cryptic species complexes: evidence from the molecular phylogeny of *Brachionus plicatilis* (Rotifera). Evolution 56:1431–1444

Gómez A, Montero-Pau J, Lunt DH, Serra M, Campillo S (2007) Persistent genetic signatures of colonization in *Brachionus manjavacas* rotifers in the Iberian Peninsula. Mol Ecol 16:3228–3240

Gribble KE, Mark Welch DB (2012) The mate recognition protein gene mediates reproductive isolation and speciation in the *Brachionus plicatilis* cryptic species complex. BMC Evol Biol 12:134

Gribble KE, Snell TW, Mark Welch DB (2011) Gene and protein structure of the mate recognition protein gene family in *Brachionus manjavacas* (Rotifera). Hydrobiologia 662:35–42

Hagiwara A, Kotani T, Snell TW, Assava-Aree M, Hirayama K (1995) Morphology, reproduction, genetics, and mating behavior of small, tropical marine *Brachionus* strains (Rotifera). J Exp Mar Biol Ecol 194:25–37

Hagiwara A, Suga K, Akazawa A, Kotani T, Sakakura Y (2007) Development of rotifer strains with useful traits for rearing fish larvae. Aquaculture 268:44–52

Hebert PDN, Cywinska A, Ball SL, DeWaard JR (2003) Biological identifications through DNA barcodes. Proc R Soc Lond B 270:313–321

Hewitt GM (1993) Postglacial distribution and species substructure: lessons from pollen, insects and hybrid zones. In: Lees DR, Edwards D (eds) Evolutionary patterns and processes. Academic Press, pp 97–103

Hey J (2001) The mind of the species problem. Trends Ecol Evol 16:326–329

Hey J (2006) On the failure of modern species concepts. Trends Ecol Evol 21:447–450

Hines HM (2008) Historical biogeography, divergence times, and diversification patterns of Bumble Bees (Hymenoptera: Apidae: Bombus). Syst Biol 57:58–75

Hwang DS, Dahms HU, Park HG, Lee JS (2013) A new intertidal Brachionus and intrageneric phylogenetic relationships among Brachionus as revealed by allometry and CO1-ITS1 gene analysis. Zoolog Stud 52:1–10

Jenkins DG, Buikema AL Jr (1998) Do similar communities develop in similar sites? A test with zooplankton structure and function. Ecol Monogr 68:421–443

Jersabek CD, De Smet WH, Fischer C, Fontaneto D, Michaloudi E, Wallace RL, Segers H (2011) List of available names in zoology, candidate part phylum Rotifera, species-group names established before 1 January 2000. International Commission on Zoological Nomenclature. http://iczn.org/lan/rotifer

Knowlton N (2000) Molecular genetic analyses of species boundaries in the sea. Hydrobiologia 420:73–90

Koste W (1978) Rotatoria. Die Rädertiere Mitteleuropas. Gebr. Borntraeger, Berlin. 673 pp, 234 plates

Kostopoulou V, Carmona MJ, Divanach P (2012) The rotifer *Brachionus plicatilis*: an emerging bio-tool for numerous applications. J Biol Res Thessaloniki 17:97–112

Lapesa S, Snell TW, Fields DM, Serra M (2002) Predatory interactions between a cyclopoid copepod and three srotifer sibling species. Freshw Biol 47:1693–1695

Lapesa S, Snell TW, Fields DM, Serra M (2004) Selective feeding of *Arctodiaptomus salinus* (Copepoda, Calanoida) on co-occurring sibling rotifer species. Freshw Biol 49:1053–1061

Losos JB (2008) Phylogenetic niche conservatism, phylogenetic signal and the relationship between phylogenetic relatedness and ecological similarity among species. Ecol Lett 11:995–1003

Lowe CD, Kemp SJ, Díaz-Avalos C, Montagnes DJS (2007) How does salinity tolerance influence the distributions of *Brachionus plicatilis* sibling species? Mar Biol 150:377–386

Lubzens E, Zmora O, Barr T (2001) Biotechnology and aquaculture of rotifer. Hydrobiologia 446/447:337–353

Malekzadeh Viayeh R, Pak Tarmani R, Rostamkhani N, Fontaneto D (2014) Diversity of the rotifer *Brachionus plicatilis* species complex (Rotifera: Monogononta) in Iran through integrative taxonomy. Zool J Linnean Soc 170:233–244

Mallet J (1995) A species definition for the modern synthesis. Trends Ecol Evol 10:294–299

May RM (1978) The dynamics and diversity of insect faunas. In: Mound LA, Walof N (eds) Diversity of insect faunas. Blackwell, Oxford, pp 188–204

Michaloudi E, Mills S, Papakostas S, Stelzer CP, Triantafyllidis A, Kappas I, Vasileiadou K, Proios K, Abatzopoulos TJ (2017) Morphological and taxonomic demarcation of Brachionus asplanchnoidis Charin within the Brachionus plicatilis cryptic species complex (Rotifera, Monogononta). Hydrobiologia 796:19–37

Mills S, Alcántara-Rodríguez JA, Ciros-Pérez J, Gómez A, Hagiwara A, Galindo KH, Jersabek CD, Malekzadeh-Viayeh R, Leasi F, Lee JS, Mark Welch DB, Papakostas S, Riss S, Segers H, Serra M, Shiel R, Smolak R, Snell TW, Stelzer CP, Tang CQ, Wallace RL, Fontaneto D, Walsh EJ (2017) Fifteen species in one: deciphering the Brachionus plicatilis species complex (Rotifera, Monogononta) through DNA taxonomy. Hydrobiologia 796:39–58. doi:10.1007/s10750-016-2725-7

Montero-Pau J, Serra M (2011) Life-cycle switching and coexistence of species with no niche differentiation. PLoS One 6:e20314

Montero-Pau J, Ramos-Rodríguez E, Serra M, Gómez A (2011) Long-term coexistence of rotifer cryptic species. PLoS One 6:e21530

Nosil P (2008) Speciation with gene flow could be common. Mol Ecol 17:2103–2106

Nosil P (2012) Ecological speciation. Oxford University Press, New York, p 304

Oogami H (1976) On the morphology of *Brachionus plicatilis*. Newsl Form Izu Branch Shizuoka Prefectural Fish Res Cent 18:2–5. (in Japanese)

Ortells R, Gómez A, Serra M (2003) Coexistence of cryptic rotifer species: ecological and genetic characterisation of *Brachionus plicatilis*. Freshw Biol 48:2194–2202

Papakostas S, Dooms S, Triantafyllidis A, Deloof D, Kappas I, Dierckens K, De Wolf T, Bossier P, Vadstein O, Kui S, Sorgeloos P, Abatzopoulos TJ (2006) Evaluation of DNA methodologies in identifying *Brachionus* species used in European hatcheries. Aquaculture 255:557–564

Papakostas S, De Wolf T, Triantafyllidis A, Vasileiadou K, Kanellis D, Cecconi P, Kappas I, Abatzopoulos TJ (2007) Follow-up of hatchery rotifer cultures with regard to their genetic identity. J Biol Res Thessaloniki 7:41–49

Papakostas S, Michaloudi E, Triantafyllidis A, Kappas I, Abatzopoulos TJ (2013) Allochronic divergence and clonal succession: two microevolutionary processes sculpturing population structure of *Brachionus* rotifers. Hydrobiologia 700:33–45

Pfenninger M, Schwenk K (2007) Cryptic animal species are homogeneously distributed among taxa and biogeographical regions. BMC Evol Biol 7:121

Pigliucci M (2003) Species as family resemblance concepts: the (dis-)solution of the species problem? BioEssays 25:596–602

Puillandre N, Lambert A, Brouillet S, Achaz G (2012) ABGD, automatic barcode gap discovery for primary species delimitation. Mol Ecol 21:1864–1877

Rico-Martinez R, Snell TW (1995) Mating behavior and mate recognition pheromone blocking of male receptors in *Brachionus plicatilis* Müller (Rotifera). Hydrobiologia 313/314:105–110

Rossberg AG, Rogers T, McKane AJ (2013) Are there species smaller than 1 mm? Proc R Soc Lond B 280:1248

Rumengan IFM, Kayano H, Hirayama K (1991) Karyotypes of S-type and L-type rotifers *Brachionus plicatilis* Müller. J Exp Mar Biol Ecol 154:171–176

Segers H (1995) Nomenclatural consequences of some recent studies on *Brachionus plicatilis* (Rotifera, Brachionidae). Hydrobiologia 313/314:121–122

Serra M, Miracle MR (1983) Biometric analysis of *Brachionus plicatilis* ecotypes from Spanish lagoons. Hydrobiologia 104:279–291

Snell TW, Carrillo K (1984) Body size variation among strains of the rotifer *Brachionus plicatilis*. Aquaculture 37:359–367

Snell TW, Stelzer C-P (2005) Removal of surface glycoproteins and transfer among *Brachionus* species. Hydrobiologia 546:267–274

Snell TW, Rico-Martinez R, Kelly LN, Battle TE (1995) Identification of a sex pheromone from a rotifer. Mar Biol 123:347–353

Snell TW, Shearer TL, Smith HA, Kubanek J, Gribble KE, Mark Welch DB (2009) Genetic determinants of mate recognition in *Brachionus manjavacas* (Rotifera). BMC Biol 7:60

Soberón J, Nakamura M (2009) Niches and distributional areas: concepts, methods, and assumptions. Proc Natl Acad Sci USA 106:19644–19650

Stelzer C-P, Riss S, Stadler P (2011) Genome size evolution at the speciation level: the cryptic species complex *Brachionus plicatilis* (Rotifera). BMC Evol Biol 11:90

Suatoni E, Vicario S, Rice S, Snell TW, Caccone A (2006) An analysis of species boundaries and biogeographic patterns in a cryptic species complex: the rotifer *Brachionus plicatilis*. Mol Phylogenet Evol 41:86–98

Sudzuki M (1987) Intraspecific variability of *Brachionus plicatilis*. Hydrobiologia 147:45–47

Tang CQ, Leasi F, Obertegger U, Kieneke A, Barraclough TG, Fontaneto D (2012) The widely used small subunit 18S rDNA molecule greatly underestimates true diversity in biodiversity surveys of the meiofauna. Proc Natl Acad Sci USA 109:16208–16212

Tang CQ, Obertegger U, Fontaneto D, Barraclough TG (2014) Sexual species are separated by larger genetic gaps than asexual species in rotifers. Evolution 68:2901–2916

Tautz D, Arctander P, Minelli A, Thomas RH, Vogler AP (2003) A plea for DNA taxonomy. Trends Ecol Evol 18:70–74

Templeton AR (1989) The meaning of species and speciation: a genetic perspective. In: Otte D, Endler JA (eds) Speciation and its consequences. Sinauer Associates, Sunderland, pp 3–27

Walczynska A, Serra M (2014) Species size affects hatching response to different temperature regimes in a rotifer cryptic species complex. Evol Ecol 28:131–140

Walker KF (1981) A synopsis of ecological information on the saline lake rotifer *Brachionus plicatilis* Müller 1786. Hydrobiologia 81:159–167

Weir JT, Schluter D (2007) The latitudinal gradient in recent speciation and extinction rates of birds and mammals. Science 315:1574–1576

Wiens JJ (2004a) What is speciation and how should we study it? Am Nat 163:914–923

Wiens JJ (2004b) Speciation and ecology revisited: phylogenetic niche conservatism and the origin of species. Evolution 58:193–197

Yoshimura K, Hagiwara A, Yoshimatsu T, Kitajima C (1996) Culture technology of marine rotifers and the implications for intensive culture of marine fish in Japan. Mar Freshw Res:217–222

Yúfera M (2001) Studies on *Brachionus* (Rotifera): an example of interaction between fundamental and applied research. Hydrobiologia 446/447:383–392

Zhang J, Kapli P, Pavlidis P, Stamatakis A (2013) A general species delimitation method with applications to phylogenetic placements. Bioinformatics (Oxford, England) 29:2869–2876

Part II
Live Food

Chapter 3
Mass Culture and Preservation of *Brachionus plicatilis* sp. Complex

Atsushi Hagiwara, Hee-Jin Kim, and Helen Marcial

Abstract The quality and quantity of rotifer production in a hatchery primarily determine the yield of larval production; therefore, considerable studies geared toward mass culture and preservation of rotifers. Rotifer mass production, attaining ultrahigh density and greater culture stability, was attained when condensed microalgal pastes (*Chlorella* and *Nannochloropsis*) were introduced and used in a closed recirculation system. The rotifer culture health status can be determined by measuring water quality, rotifer physiological conditions such as egg ratio, swimming speed, ingestion rate, and in vivo enzyme that takes 1–2 days to finish. Addition of chemicals and hormones such as γ-aminobutyric acid (GABA), porcine growth hormone, serotonin, and human chorionic gonadotropin is known to improve the health status of cultured rotifers. Mass-produced rotifers can be preserved at low temperature; these rotifers can either be directly used as feed for the larvae or as starters of rotifer culture. For long-term storage, collection and storage of resting eggs are recommended. For species which do not produce resting eggs, cryopreservation of amictic eggs is recommended.

3.1 Introduction

The rotifer *Brachionus plicatilis* species complex is being used worldwide as a live food for the initial stages of larval rearing of fish and shellfish because of their suitable size and shape. Based on morphometric characteristics, rotiferologists have classified *B. plicatilis* into three morphotypes (Fu et al. 1991): L-type (130–340 μm), S-type (100–210 μm), and SS-type (90–150 μm; see review Hagiwara et al. 1995, 2001). This classification is supported by biochemical and ecological differences

A. Hagiwara (✉) • H.-J. Kim
Graduate School of Fisheries and Environmental Sciences, Nagasaki University,
Bunkyo 1-14, Nagasaki 852-8521, Japan
e-mail: hagiwara@nagasaki-u.ac.jp

H. Marcial
Center for International Collaborative Research, Nagasaki University,
Bunkyo 1-14, Nagasaki 852-8521, Japan

A. Hagiwara, T. Yoshinaga (eds.), *Rotifers*, Fisheries Science Series,
DOI 10.1007/978-981-10-5635-2_3

(Snell 1998), including mating behavior (based on 29-kDa glycoprotein sex pheromone; see Snell et al. 1995), hence, assigned a name to distinguish these different species as *B. plicatilis* for L-type, *B. ibericus* for S-type, and *B. rotundiformis* for SS-type (Kotani et al. 2005). However, this physiology-based classification is considered a difficult approach with weak boundary species (Kotani et al. 1997); thus, molecular phylogeny was adapted to classify rotifer species complex. Several cryptic species were found within *B. plicatilis* sp. complex using ITS1and COI sequences (Gómez et al. 2002), and to date, 15 different species are classified (Mills et al. 2016). Among rotifers, this taxon is well studied because of its ease in culture and high reproductive performance and thus commonly used in fish and crustacean hatcheries (Hagiwara et al. 2007). The quality and quantity of rotifer production in a hatchery primarily determine the yield of larval production; therefore, considerable attention and researches are being done in the mass culture and preservation of this species complex (Hagiwara 1994; Lubzens and Zmora 2003).

Mass production of *B. plicatilis* sp. complex was initiated in the early 1960s with the use of *Nannochloropsis oculata* as diet (Hirata 1964). Thereafter, rotifer mass culture trials were conducted using various diets such as baker's yeast, marine yeast, bacteria, omega yeast, and *Tetraselmis tetrathele* (reviewed by Nagata and Hirata 1986). Rotifer mass production has been systematized with the combination of improved diet formulation (Yoshimatsu and Hossain 2014), and rotifer densities reached to 2.3×10^4 ind/mL (Dhert et al. 2001). However, in spite of these advancements, fresh phytoplankton and/or baker's yeast is still generally used for the mass culture of rotifers worldwide. The use of baker's yeast is beneficial for aquaculturists since it lowers the cost for rotifer production. However, rotifer cultures fed yeast are less stable owing to the rapid decline of water quality. Moreover, rotifers fed yeast need further nutritional enrichment before being fed to fish larvae; therefore, phytoplankton remains the ideal food for rotifer cultures (Lubzens et al. 1995a). However, culturing of phytoplankton requires considerable space in a hatchery. Condensed microalgal paste was found to be the ideal food for ultrahigh production of rotifers. During the late 1980s, a phytoplankton industry has developed new products such as condensed microalgal paste (*Chlorella* and *Nannochloropsis*), which enabled culturists to mass culture rotifers at a high density as much as 1.6×10^5 ind/mL (Lubzens et al. 1995a; Yoshimura et al. 1996, 2003; Hagiwara and Kuwada 2004) with higher stability of culture. Furthermore, it was the most convenient, since it could be stored at low temperature (refrigerator or freezer) for at least 2 months or even longer without significant loss of its essential fatty acid content and maintaining its nutrient composition in cultured rotifers (Lubzens et al. 1995a; Welladsen et al. 2014). However, condensed microalgal pastes are not commercially available in developing countries because of higher price. Recently, detritus from macroalga, *Ulva pertusa*, was shown to sustain high population growth and enhance the nutritional value of *B. plicatilis* (Yin et al. 2013). Detrital macroalgae can also be stored and collected from the wild; thus, it is considered economical in large-scale production of rotifers.

In this chapter, various culture methods commonly used for mass culture of clonal rotifer *B. plicatilis* sp. complex are discussed. Since the major problem encountered by aquaculturists is the unpredictable culture collapse, some of the techniques used to diagnose and treat rotifer culture are presented. To address the inadequate supply of larval food, some preservation methods that have been proposed by many researchers are also discussed.

3.2 Culture Methods

3.2.1 Batch Culture

Batch culture is a common method for rotifer mass production in marine fish hatcheries (Fig. 3.1a). The culture maintains a constant volume with an increasing rotifer density or a constant rotifer density by increasing the culture volume (Dhert et al. 2001). When the desired density is achieved, the culture is entirely harvested at once by draining the culture medium and collecting the rotifers in a net. The harvested rotifers are used to feed the larvae, and the remaining is used as the inoculum for the next culture (Lubzens 1987; Hino 1993; Dhert et al. 2001). The size of culture vessels is flexible: 500–1000 L for plastic tanks or up to 10 ton for concrete tanks. In case of S-type rotifers, the obtainable densities at harvest time are over 1000 ind/mL when fed with *Nannochloropsis oculata* and/or backer's yeast (Hino 1993) and 600 ind/mL when fed with artificial diet (Culture Selco®) (Suantika et al. 2000). This method is used in many countries; however, it has many disadvantages, including low efficiency in terms of labor and utilization of infrastructure, low production yield, unstable and unpredictable culture, and high costs of operation (Dhert et al. 2001).

3.2.2 Semicontinuous Culture

The semicontinuous culture method is also known as "thinning out" culture, because the rotifer density is maintained constant by periodic harvesting (Hino 1993). The size of the culture tank is usually larger (usually between 20 and 480 tons) than that used in batch culture. The initial density of rotifers inoculated into the system varies between 50 and 200 ind/mL and might reach 300 to over 1000 ind/mL in 3–7 days when fed fresh microalgae and/or baker's yeast (Dhert et al. 2001). Unlike in the batch culture method, in this system, a fixed fraction of culture water (which contains rotifers and residual food) is harvested at regular intervals and replaced by an equal quantity of fresh culture medium (Fig.3.1b). The dilution rate determines the rotifer population dynamics. Monod kinetics and related mathematical concepts are used to model the quasi-steady state of periodically diluted cultures used in this system (Schlüter et al. 1987). Navarro and Yufera (1998) used the quasi-steady state method and freeze-dried microalgae and used dilution rates of 0.3/day and 0.2/day, which resulted in the best production (mg rotifer/day) and food conversion efficiency (mg rotifer developed/mg microalgae consumed) for L- and S-type rotifers, respectively.

3.2.3 High-Density Culture

An intensive mass culture system for rotifers was developed by Japanese scientists in the late 1990s. In the culture trial with the use of concentrated freshwater *Chlorella* diet, an ultrahigh S-type rotifer density ranging from 10,000 to 30,000 ind/mL was obtained (Yoshimura et al. 1996). Under this batch, culture method, culture instability due to ammonia accumulation, presence of bacteria/protozoa, and food shortage or oxygen decline, however, are often encountered (Hagiwara and Kuwada 2004). To solve these problems, Yoshimura et al. (2003) employed several remedies including oxygen gas supplementation, regulation of pH (adjusted to 7), and a filtering equipment which prevents particulate organic matter, debris, and bacteria from clogging the collection net during harvest. In addition, Fu et al. (1997) introduced an automatic *continuous culture system* as the solution to maintain culture stability, as well as to reduce labor and create more space in the hatchery (Fig. 3.1c). This system consists of a filtration unit, culture unit, and harvest unit. Filtered water and food are continuously supplied into the rotifer culture tank at a predetermined rate, and the same amount of culture water is transferred into the harvest tank to obtain a significant rotifer biomass. By using this system, Fu et al. (1997) successfully mass produced about 2.1 billion rotifers/day for S-type in a 1-m^3 tank with densities ranging from 3000 to 6000 ind/mL and about 0.17 billion rotifers/day for L-type in a 500-L tank with densities ranging from 1100 to 2200 ind/mL. The longest duration of their culture was over 110 days. Hagiwara and Kuwada (2004) described continuous rotifer culture system using a large tank (20 ton), which uses facilities in old hatcheries. Commercially available concentrated freshwater *Chlorella vulgaris* was used as feed for both species. In order to achieve high-density rotifers, a *closed recirculation culture system* (Fig. 3.1d) was lastly performed. This system comprised of protein skimmers, novel filters, ozone, and an addition of sodium hydroxymethanesulfonate to neutralize ammonia in the tank. With these modifications, ultrahigh rotifer densities (over 5000 ind/mL) were maintained for longer culture period (up to 30 days) without compromising the water quality (Suantika et al. 2000, 2001; Bentley et al. 2008). The development of a high-density rotifer culture system significantly reduced the space needed at the hatchery for live food production, allowing the production of sufficient amount of rotifers even in small tanks, improved the water quality, and decreased the harmful bacterial load of the culture; thus, this system was significantly economical (Suantika et al. 2003).

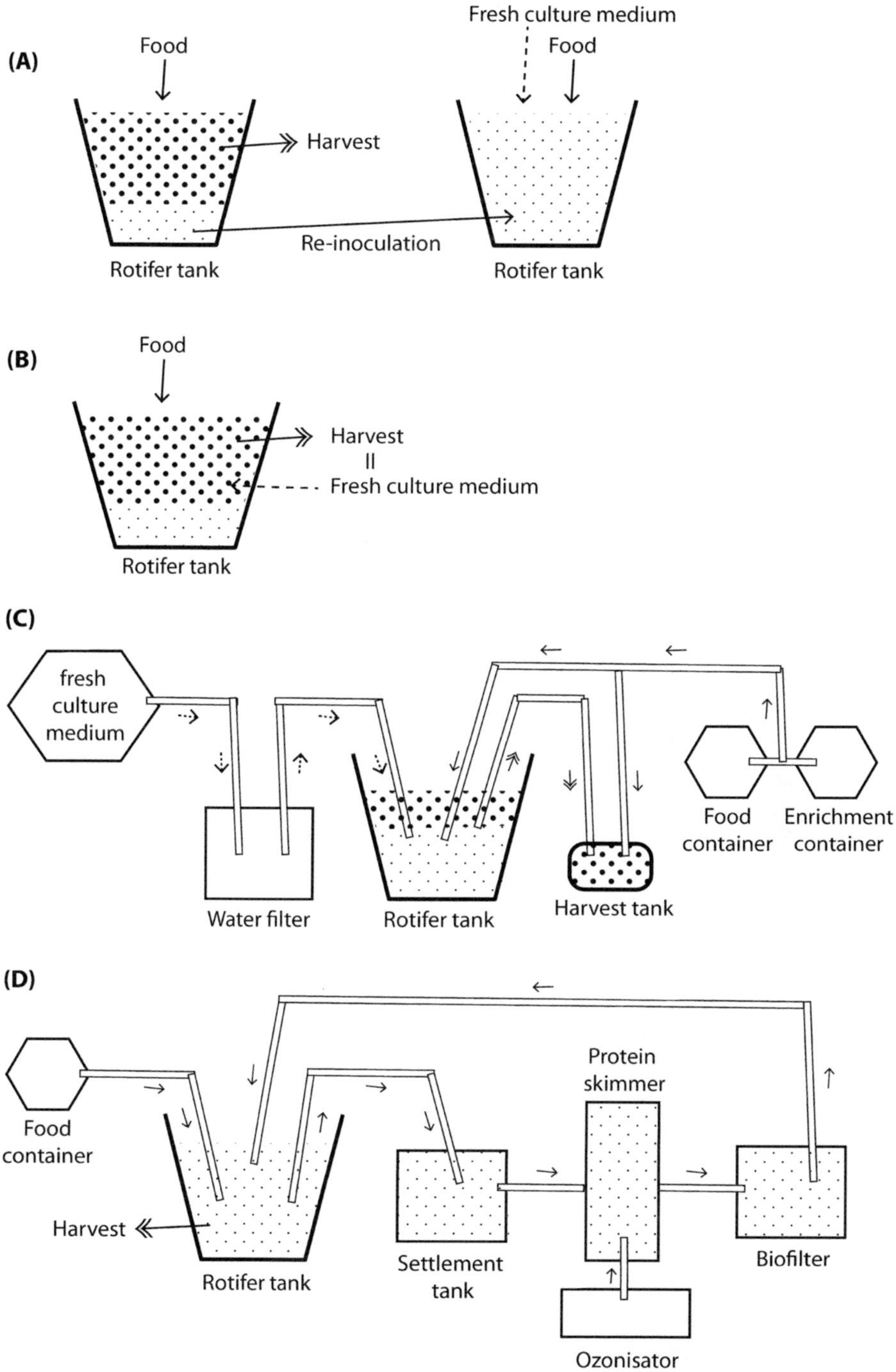

Fig. 3.1 Various mass culture systems of rotifer *Brachionus plicatilis* sp. complex. (**a**) Batch culture, (**b**) semicontinuous culture, and high-density culture including (**c**) automatic continuous and (**d**) closed recirculation systems. Normal, *dot*, and *double head arrows* indicate major inflows of materials, fresh medium, and rotifer harvest line, respectively

Table 3.1 Effects of chemicals on reproduction (i.e., population growth and mixis induction) and body size of the rotifer *B. plicatilis* sensu stricto

Chemicals (concentrations affect reproduction)	Population growth	Mixis induction	Body size
γ-Aminobutyric acid (50 mg/L)	++	++	+
Porcine growth hormone (0.025 I.U./mL)	++	++	n.e.
Serotonin (0.05 and 5 mg/L)	+	++	n.e.
Human chorionic gonadotropin (2.5 I.U./mL)	+	n.e.	–
Juvenile hormone (0.05 and 0.5 mg/L)	n.e.	++	++
Estradiol-17β (50 mg/L)	n.e.	+	–
20-hydroxyecdysone (0.05 mg/L)	n.e.	+	–

Adopted from Hagiwara et al. (2001) and Hagiwara and Kuwada (2004)
++, increased phenomenon ($P < 0.01$); +, increased phenomenon ($P < 0.05$); –, decreased phenomenon ($P < 0.05$); n.e., no effect ($P > 0.05$).

3.3 Culture Diagnosis and Treatment

Rotifers have been successfully mass produced; however, maintenance of the culture for a long time remains a critical unresolved problem. Several factors, including decreased feeding activity, protozoan contamination, and poor water quality, caused rotifer cultures to collapse (Yu and Hirayama 1986; Jung et al. 1997). The water quality in the rotifer culture can be assessed by measuring pH, ammonia level, and viscosity (Snell et al. 1987; Hagiwara et al. 1998). The reported critical pH and unionize ammonia concentrations are at 7 (Yoshimura et al. 1996) and at less than 2.1 ppm (Yu and Hirayama 1986), respectively. The viscosity of rotifer culture medium increases with the accumulation of dissolve organic substances, and the higher viscosity causes a decline in the rotifer population (Araujo et al. 2001). It is therefore necessary to continuously monitor rotifer cultures in order to determine their health status. Several methods are recommended to assess the physiological status of cultured rotifers, as well as to predict culture collapse (for review see Hagiwara and Kuwada 2004). Egg ratio (the number of parthenogenetic eggs per female) was used by Snell et al. (1987); however, this method required 1–2 days for the assessment and was not a sensitive end point for culture diagnosis. Swimming speed (Snell et al. 1987; Janssen et al. 1994) and ingestion rate (Ferrando et al. 1993; Juchelka and Snell 1994) were also found to be sensitive indicators to detect stress in rotifer culture. These two characteristics can be detected easily and rapidly (require around 1–2 h). Furthermore, in vivo enzyme activity test, which is considered to be the most sensitive method (Araujo et al. 2001), is another way to assess the health of rotifer cultures; however, it requires instruments such as a computer, fluorometer, and an image analyzer, which increases the cost for operation.

Several treatment methods have been suggested to improve rotifer culture and prevent collapse (Table 3.1). Araujo and Hagiwara (2005) found that the addition of γ-aminobutyric acid (GABA) can improve the health condition of rotifers when they are exposed to stressful environment (e.g., increase of unionized ammonia and protozoa contamination). Addition of GABA during nutritional enrichment culture is also shown to improve survival and swimming activity of rotifers (Gallardo et al. 2001). In addition, the supplementation of porcine growth hormone significantly enhances the rotifer population growth when environmental stressors are low, such as under optimal food conditions with a low concentration of free ammonia (Gallardo et al. 1999). Serotonin and human chorionic gonadotropin are also effective to increase the rotifer population growth. For the improvement of rotifer sexual reproduction, in addition to GABA, serotonin can also be applied under low food conditions. On the other hand, juvenile hormone increases mixis production under optimal and suboptimal food conditions. To apply the mentioned chemicals into rotifer mass cultures, the following issues have to be dealt with: (1) several chemicals also affect body size of rotifers (Gallardo et al. 1997) and unknown influences remain, (2) chemical-treated rotifers would influence fish larvae, and (3) effluents containing these hormones would have effects on indigenous species in receiving waters (Hagiwara et al. 2001).

3.4 Preservation

Despite the progress of successful establishment of mass culture of rotifers, diagnosis of culture status, treatment of culture for recovery, and methods to preserve rotifers for either feeding fish stocks or starting a new culture are necessary. Mass preservation is essential because of the usual problem of culture collapse. Rotifers can be preserved for a short time (few days to 4 weeks) at low temperatures (between −2 and 8 °C) and for long term (years) by producing and collecting resting eggs.

Rotifers preserved at low temperature can either be directly used as feed for the larvae or as starters of rotifer culture (Assavaaree et al. 2001; reviewed by Hagiwara et al. (2001)). However, *B. plicatilis* sp. complex respond differently to various conditions during preservation. Assavaaree et al. (2001) found that *B. plicatilis* s. s. were more resistant to low-temperature (4 °C) preservation than *B. rotundiformis* (Fig. 3.2). They also found that exchange of culture media during the incubation at low temperature is necessary to maintain the viability of the rotifers. The rotifer survival at low temperature is also significantly related to the rotifer culture conditions, e.g., food and salinity before exposure to low temperature (Lubzens et al. 1990), where syntheses of specific proteins, including 94 kD and HSP 60, are being synthesized (Lubzens et al. 1995b). The recovery rate of preserved *B. plicatilis* for 14 days at density of 20,000 ind/mL was about 50%. The S- and SS-type *B. rotundiformis* strains were comparatively less resistant to low-temperature preservation but can be improved by GABA treatment (Assavaaree and Hagiwara 2011).

For long-term storage, resting eggs of rotifers can be produced and hatched when needed. The resting egg is a notable feature of rotifer life cycle and is the end product of sexual reproduction which is resistant to harsh environment. The advantages of rotifer resting eggs are highlighted for preservation (reviewed by Hagiwara and Hirayama 1993; Hagiwara et al. 1997). These eggs can be stored for more than

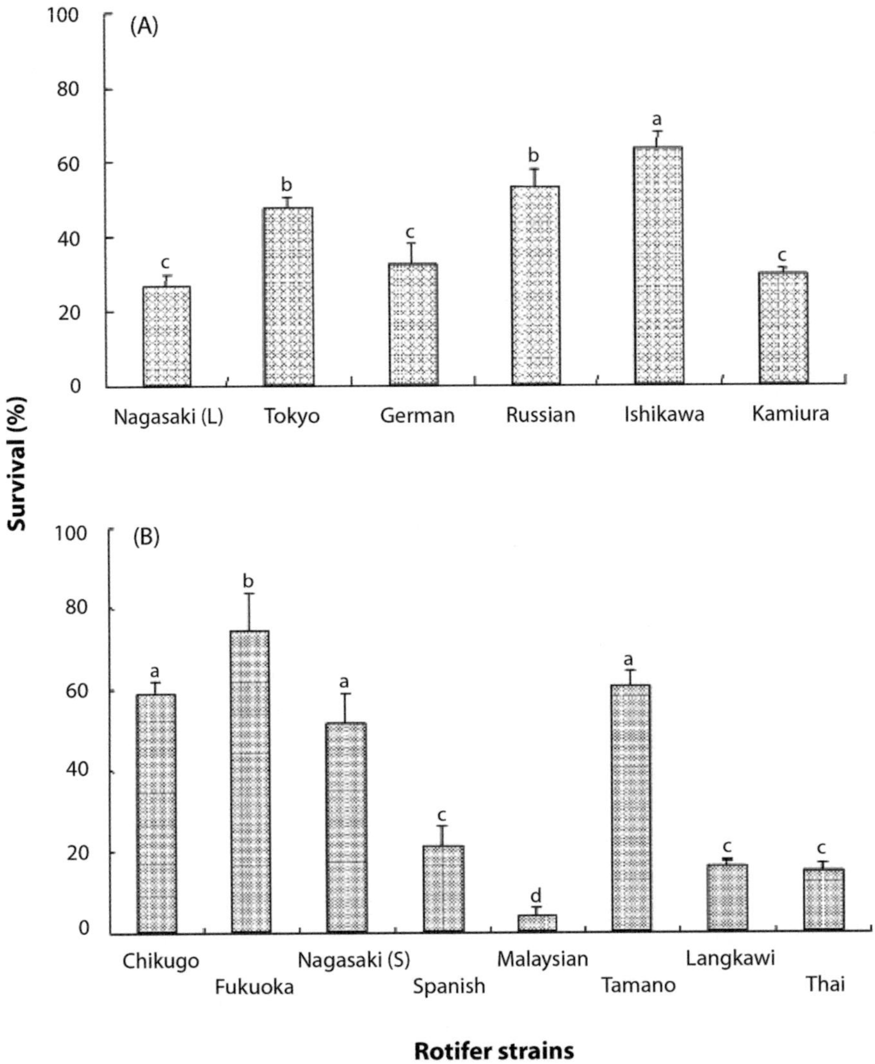

Fig. 3.2 Percent survival after preservation at 4 °C. (**a**) Survival of 6 L-type strains on day 30 and (**b**) survival of 8 S- and SS-type strains on day 5. Columns and vertical bars indicate mean rotifer survival of three replicates and standard deviation, respectively. Results of Tukey test were presented (a>b>c, $P < 0.05$) (Adopted from Hagiwara et al. 2001)

20 years in sterilized seawater under complete darkness at 5 °C. The hatching rates of dried resting egg gradually decrease, but those preserved by canning at an atmospheric pressure of 48–61 kPa after lyophilization (at −30 °C) can be maintained for up to 6 months (Hagiwara et al. 1997). The resting egg production, however, is an obstacle for mass production of rotifers because of the appearance of mictic females

which cannot contribute to population growth. Resting eggs, however, could be a by-product during hatchery off-season operation and then can be hatched with the same manner as *Artemia* cysts when needed. However, even with these merits, not all species of *B. plicatilis* sp. complex produce resting eggs, and hatchlings from resting eggs would have different reproductive characteristics from their parent because of gene recombination during resting egg formation. As solution to these issues, other methods such as cryopreservation of amictic eggs were tested (Toledo and Kurokura 1990; Toledo et al. 1991; Lubzens et al. 2001). With this method, rotifers showed around 50% of survival rate. This method is recommended and is useful for the maintenance of a certain strain with desired morphological and genetic characteristics. Nevertheless, there are several drawbacks to this technique such as practical only in small-scale and low population growth after thawing (Toledo et al. 1991).

References

Assavaaree M, Hagiwara A (2011) Effect of gamma-aminobutyric acid and porcine growth hormone on survival of the euryhaline rotifers *Brachionus plicatilis* sp. complex preserved at low temperature. Fish Sci 77:599–605

Assavaaree M, Hagiwara A, Ide K, Maruyama K, Lubzens E (2001) Low-temperature preservation (at 4°C) of marine rotifer *Brachionus*. Aquac Res 32:29–39

Bentley CD, Carroll PM, Watanabe WO (2008) Intensive rotifer production in a pilot-scale continuous culture recirculating system using nonviable microalgae and an ammonia neutralizer. J World Aquacult Soc 39:625–635

de Araujo AB, Hagiwara A (2005) Screening methods for improving rotifer culture quality. Hydrobiologia 546:553–558

de Araujo AB, Hagiwara A, Snell TW (2001) Effect of unionized ammonia, viscosity and protozoan contamination on reproduction and enzyme activity of the rotifer *Brachionus rotundiformis*. Hydrobiologia 446(447):363–368

Dhert P, Rombaut G, Suantika G, Sorgeloos P (2001) Advancement of rotifer culture and manipulation techniques in Europe. Aquaculture 200:129–146

Ferrando MD, Janssen CR, Persoone G (1993) Ecotoxicological studies with the freshwater rotifer *Brachionus calyciflorus*. III. The effects of chemicals on feeding behavior. Ecotoxicol Environ Saf 26:1–9

Fu Y, Hirayama K, Natsukari Y (1991) Genetic divergence between S and L type strains of the rotifer Brachionus plicatilis O. F. Müller. J Exp Mar Biol Ecol 151:43–56

Fu Y, Hada A, Yamashita T, Yoshida Y, Hino A (1997) Development of a continuous culture system for stable mass production of the marine rotifer *Brachionus*. Hydrobiologia 358:145–151

Gallardo WG, Hagiwara A, Tomita Y, Soyano K, Snell TW (1997) Effect of some vertebrate and invertebrate hormones on the population growth, mictic female production, and body size of the marine rotifer *Brachionus plicatilis* Müller. Hydrobiologia 358:113–120

Gallardo WG, Hagiwara A, Tomita Y, Snell TW (1999) Effect of growth hormone and gamma-aminobutyric acid on *Brachionus plicatilis* (Rotifera) reproduction at low food or high ammonia levels. J Exp Mar Biol Ecol 240:179–191

Gallardo WG, Hagiwara A, Snell TW (2001) Use of GABA to enhance rotifer reproduction in enrichment culture. Aquac Res 32:243–246

Gómez A, Serra M, Carvalho GR, Lunt DH (2002) Speciation in ancient cryptic species complexes: evidence from the molecular phylogeny of *Brachionus plicatilis* (Rotifera). Evolution 56:1431–1444

Hagiwara A (1994) Practical use of rotifer cysts. Israel J Aquac Bamidgeh 46:13–21

Hagiwara A, Hirayama K (1993) Preservation of rotifers and its application in the finfish hatchery. In: Lee CS, Su MS, Liao IC (Eds.) Finfish Hatchery in Asia '91. TML Conference Proceedings 3:61–71. Tunkang Marine Laboratory, Taiwan Fisheries Research Institute, Tungkang, Pingtung, Taiwan

Hagiwara A, Kuwada H (2004) Current status of live food culture in Japan. The Second Hatchery Feeds and Technology Workshop. Sydney, Australia, pp. 24–33

Hagiwara A, Kotani T, Snell TW, Assavaaree M, Hirayama K (1995) Morphology, reproduction, genetics and mating behavior of small, tropical marine rotifer *Brachionus* strains (Rotifera). J Exp Mar Biol Ecol 194:25–37

Hagiwara A, Balompapueng MD, Munuswamy N, Hirayama K (1997) Mass production and preservation of the resting eggs of the euryhaline rotifer *Brachionus plicatilis* and *B. rotundiformis*. Aquaculture 155:223–230

Hagiwara A, Yamamiya N, de Araujo AB (1998) Effect of water viscosity of the population growth of the rotifer *Brachionus plicatilis* Muller. Hydrobiologia 387(388):489–494

Hagiwara A, Gallardo WG, Assavaaree M, Kotani T, Araujo AB (2001) Live food production in Japan: recent progress and future aspects. Aquaculture 200:111–127

Hagiwara A, Suga K, Akazawa A, Kotani T, Sakakura Y (2007) Development of rotifer strains with useful traits for rearing fish larvae. Aquaculture 268:44–52

Hino A (1993) Present culture systems of the rotifer (*Brachionus plicatilis*) and the function of micro-organisms. TML Conf Proc 3:51–59

Hirata H (1964) Culture of live food at Yashima station. Saibai Gyogyo News 2:4

Janssen RC, Ferrando DM, Persoone G (1994) Ecotoxicological studies with the freshwater rotifer *Brachionus calyciflorus* IV. Rotifer behavior as a sensitive and rapid sublethal test criterion. Ecotoxicol Environ Saf 28:244–255

Juchelka CM, Snell TW (1994) Rapid toxicity assessment using rotifer ingestion rate. Arch Environ Contam Toxicol 26:549–554

Jung M-M, Hagiwara A, Hirayama K (1997) Interspecific relations in the marine rotifer microcosm. Hydrobiologia 358:121–126

Kotani T, Hagiwara A, Snell TW (1997) Genetic variations among marine rotifer strains and function of mate recognition pheromone (MRP). Hydrobiologia 358:105–112

Kotani T, Hagiwara A, Snell TW, Serra M (2005) Euryhaline *Brachionus* strains (Rotifera) from tropical habitats: morphology and allozyme patterns. Hydrobiologia 546:161–167

Lubzens E (1987) Raising rotifers for use in aquaculture. Hydrobiologia 147:245–255

Lubzens E, Zmora O (2003) Production and nutritional value of rotifers, 17–64 pp, 300–303 pp. In; Live feeds in marine aquaculture. Eds: JG Støttrup and LA McEvoy, Balckwell Science Ltd, 318 pp

Lubzens E, Kolodny G, Perry B, Galai N, Sheshinski R, Wax Y (1990) Factors affecting survival of rotifers (*Brachionus plicatilis* O. F. Müller) at 4°C. Aquaculture 91:23–47

Lubzens E, Gibson O, Zmora O, Sukenik A (1995a) Potential advantages of frozen algae (*Nannochloropsis* sp.) for rotifer (*Brachionus plicatilis*) culture. Aquaculture 133:295–309

Lubzens E, Rankevich D, Kolodny G, Gibson O, Cohen A, Khayat M (1995b) Physiological adaptations in the survival of rotifers (*Brachionus plicatilis*, O.F. Müller) at low temperatures. Hydrobiologia 313:175–183

Lubzens E, Zmora O, Barr Y (2001) Biotechnology and aquaculture of rotifers. Hydrobiologia 446:337–353

Mills S, Alcántara-Rodríguez A, Ciros-Pérez J, Gómez A, Hagiwara A, Galindo KH, Jersabek CD, Malekzadeh-Viayeh R, Leasi F, Lee J-S, Welch DBM, Papakostas S, Riss S, Segers H, Serra M, Shiel R, Smolak R, Snell TW, Stelzer CP, Tang CQ, Wallace RL, Fontaneto D, Walsh EJ (2016) Fifteen species in one: deciphering the *Brachionus plicatilis* species complex (Rotifera, Monogononta) through DNA taxonomy. Hydrobiologia. doi:10.1007/s10750-016-2725-7

Nagata WD, Hirata H (1986) Mariculture in Japan: past, present, and future prospective. Mini Rev Data File Res 4:1–38

Navarro N, Yufera M (1998) Population dynamics of rotifers *Brachionus plicatilis* and *Brachionus rotundiformis* in semi-continuous culture fed freeze-dried microalgae: influence of dilution rate. Aquaculture 166:297–309

Schlüter M, Soeder CJ, Groeneweg J (1987) Growth and food conversion of *Brachionus rubens* in continuous culture. J Plankton Res 9:761–783

Snell TW (1998) Chemical ecology of rotifers. Hydrobiologia 387:267–276

Snell TW, Childress MJ, Boyer EM, Hoff FH (1987) Assessing the status of rotifer mass cultures. J World Aquacult Soc 18:270–277

Snell TW, Rico-Martinez R, Kelly LN, Battle TE (1995) Identification of a sex pheromone from a rotifer. Mar Biol 123:347–353

Suantika G, Dhert P, Nurhudah M, Sorgeloos P (2000) High-density production of the rotifer *Brachionus plicatilis* in a recirculation system: consideration of water quality, zootechnical and nutritional aspects. Aquac Eng 21:201–213

Suantika G, Dhert P, Rombaut G, Vandenberghe J, de Wolf T, Sorgeloos P (2001) The use of ozone in a high density recirculation system for rotifers. Aquaculture 201:35–49

Suantika G, Dhert P, Sweetman E, O'Brien E, Sorgeloos P (2003) Technical and economical feasibility of a rotifer recirculation system. Aquaculture 227:173–189

Toledo JD, Kurokura H (1990) Cryopreservation of the euryhaline rotifer *Brachionus plicatilis* embryos. Aquaculture 91:385–394

Toledo JD, Kurokura H, Nakagawa H (1991) Cryopreservation of different strains of the euryhaline rotifer *Brachionus plicatilis* embryos. Nippon Suisan Gakkaishi 57:1347–1350

Welladsen H, Kent M, Mangott A, Li Y (2014) Shelf-life assessment of microalgae concentrates: effect of cold preservation on microalgal nutrition profiles. Aquaculture 430:241–247

Yin XW, Min WW, Lin HJ, Chen W (2013) Population dynamics, protein content, and lipid composition of *Brachionus plicatilis* fed artificial macroalgal detritus and *Nannochloropsis* sp. diets. Aquaculture 380-383:62–69

Yoshimatsu T, Hossain MA (2014) Recent advances in the high-density rotifer culture in Japan. Aquac Int 22:1587–1603

Yoshimura K, Hagiwara A, Yoshimatsu T, Kitajima C (1996) Culture technology of marine rotifers and the implications for intensive culture of marine fish in Japan. Mar Freshw Res 47:217–222

Yoshimura K, Tanaka K, Yoshimatsu T (2003) A novel culture system for the ultra-high-density production of the rotifer, *Brachionus rotundiformis*—a preliminary report. Aquaculture 227:165–172

Yu J-P, Hirayama K (1986) The effect of un-ionized ammonia on the population growth of the rotifer in mass culture. Nippon Suisan Gakkaishi 52:1509–1513

Chapter 4
Enrichment of Rotifers and Its Effect on the Growth and Survival of Fish Larvae

Tomonari Kotani

Abstract In order to improve the survival and growth of aquacultured finfish larvae, nutritional enrichment of rotifers is necessary. In the development of methodologies for material and process, *n*-3HUFAs especially EPA and DHA were the main focused nutrients for enrichment. Although the addition of those fatty acids to rotifer was conducted in vitro, the enrichment of DHA or EPA to phospholipid was suggested recently, as well as the optimum ratio among DHA, EPA, and arachidonic acid. On the other hand, other nutrients, e.g., taurine, vitamin A, and some minerals, have also been suggested to enrich rotifers. These mean that not only specific nutrients but also the balance among nutrients is important to enrich rotifers nutritionally and improve the performance of cultured finfishes. Moreover, the influence by rotifer culture method and protocol and the development of enrichment method should be taken into consideration for improving the efficiency of enrichment. Future studies should investigate whether rotifers can acquire all these nutrients from the balanced nutrition.

4.1 Introduction

Euryhaline rotifer *Brachionus* has been used as food for finfish and shellfish larvae since Ito (1960) suggested the usage of them as feed for marine fish larvae. In Japan, Hirata (1964) introduced the marine microalga *Nannochloropsis oculata* (formally called as marine *Chlorella*) as the optimal feed for rotifer. Because of the large space required for live microalgal production, baker's yeast was used to address this problem. However, the nutrition status of rotifers fed baker's yeast was reported to be poor and could not ensure perfect growth and survival of fish larvae (Watanabe et al. 1978a, b); thus, nutritional enrichment of rotifers was required. At present, rotifers are not fed to larvae without nutritional enrichment. Although such enrichment mainly focused on fatty acids, especially *n*-3 highly unsaturated fatty acids

T. Kotani (✉)
Faculty of Fisheries, Kagoshima University, Kagoshima, Japan
e-mail: kotani@fish.kagoshima-u.ac.jp

A. Hagiwara, T. Yoshinaga (eds.), *Rotifers*, Fisheries Science Series,
DOI 10.1007/978-981-10-5635-2_4

(*n*-3 HUFAs), other essential amino acids and minerals were also included in the enrichment process (Chen et al. 2004; Matsumoto et al. 2009; Penglase et al. 2011).

In this chapter, the history of enrichment methods, specific nutritional elements, and their effects on each fish species are discussed.

4.2 Background

4.2.1 Necessity of Enrichment

Finfish larvae open the mouth when or after their yolk is absorbed or disappears. They usually started the exogenous feeding. Generally, the function of gastric gland of marine finfish larvae is not sufficient for digesting various nutrients when they open their mouth (Dabrowski 1984; Kolkovski 2001). In the natural field, the larvae fed on the plankton or the detritus, and these feeds are adequate for the digestion ability of larvae (Hunter 1981; Dabrowski 1984; Sutton and Bowen 1994; Yako et al. 1996). Finfish larvae usually require various nutrients (Kjørsvik et al. 2011; Hamre et al. 2013). Previous studies reported that phospholipid, *n*-3HUFA, taurine, vitamin C, and fat-soluble vitamins are important for the survival, growth, and normal development of larvae. These essential nutrients should be derived from exogenous feeding (Hamre et al. 2013).

Before the mouth opening or the yolk exhaustion, larvae derived nutrients from the yolk. If they cannot take some food after mouth opening, they consume their body content (Yúfera and Darias 2007; Jaroszewska and Dabrowski 2011). However, larvae of many finfish species cannot digest the artificial feed, even if they are in a form of microparticle diet (Watanabe and Kiron 1994). Although digestible artificial feeds have been developed recently, and some studies attempted to use them, many hatcheries still depend on rotifers as first feed for larvae in spite of little or absence of nutrients in rotifers. Therefore, rotifers have to be nutritionally enriched in order to feed to marine finfish larvae.

4.2.2 Noticing the Necessity of Nutritional Enrichment

In Japanese larviculture activities, *N. oculata* was used as feed for rotifers until the mid-1990s. However, the cultivation of *N. oculata* was not sufficient to meet the food requirement for rotifer culture. The fluctuation in *N. oculata* quantity was compensated by using a mixture of baker's yeast and *N. oculata* (Hirata and Mori 1967). Subsequently, only baker's yeast was used for rotifer cultures. Unfortunately, fish larvae fed on rotifers cultured using baker's yeast showed reduced activity and malformations, which in turn decreased their growth and survival (Sumida et al. 1974; Kitajima and Koda 1976; Fukusho 1977). Kitajima et al. (1979) reported that feeding rotifers fed *N. oculata* after the introduction of baker's yeast improved the growth and survival of red sea bream larvae. This was because rotifers fed

Table 4.1 Percentages of fatty acid composition in total lipid included in rotifers enriched or cultured with various materials

Type of material	*Nannochloropsis oculata*	*Chlorella vulgaris*	Yeast	Oil-supplemented yeast	*Schizochytrium* sp.
Fatty acid composition (%)					
14:0	4.0	1.1	2.5	2.0	16.2
16:0	14.4	13.6	7.1	11.1	24.8
16:1	20.4	4.6	26.5	10.8	9.9
18:0	2.2	2.4	4.3	2.3	2.6
18:1 n-9	10.5	31.5	29.1	23.0	2.3
18:2 n-6	4.7	25.7	6.9	2.1	8.0
18:3 n-3	0.1	1.8	0.2	0.8	1.4
20:4 n-6	4.1	1.5	0.9	0.7	1.4
20:5 n-3	27.7		1.4	11.9	0.3
22:6 n-3				9.0	18.3
	Watanabe et al. (1978a, b)	Kotani et al. (2013)	Imada et al. (1979)	Imada et al. (1979)	Barclay and Zeller (1997)

Table 4.2 Percentages of fatty acid composition in total lipid included in various materials to enrich or culture

Type of material	*Nannochloropsis oculata*	*Chlorella vulgaris*	Yeast	Oil-supplemented yeast	*Schizochytrium* sp.
Fatty acid composition (%)					
14:0			0.3	4.1	11.0
16:0	14.0	13.9	8.3	13.4	38.5
16:1	26.0	5.7	38.2	6.6	7.3
18:0	0.0	3.1	4.1	2.4	1.1
18:1 n-9	2.6	2.2	43.9	16.4	4.1
18:2 n-6	4.5	25.3	2.8	1.1	
18:3 n-3	0.0	24.2	0.5	0.8	
20:4 n-6	6.6	0.0		3.0	
20:5 n-3	40.0	0.0		17.7	0.6
22:6 n-3	0.0	0.0		12.8	24.0
	Maruyama et al. (1997)	Maruyama et al. (1997)	Imada et al. (1979)	Imada et al. (1979)	Barclay and Zeller (1997)

N. oculata contain *n*-3 HUFA, especially eicosapentaenoic acid (EPA), unlike ones fed baker's yeast alone (Watanabe et al. 1978a, b; Table 4.1). This indicated that *n*-3 HUFAs are an important nutrient for the normal growth and survival. Therefore, rotifers fed baker's yeast alone should be enriched with*n*-3 HUFAs. Although rotifers are mainly fed freshwater *Chlorella* in Japan, they also have to be enriched because freshwater *Chlorella* does not include *n*-3 HUFAs (Maruyama et al. 1997; Table 4.2). Since the 1980s, the nutritional enrichment has become a popular method in larval fish rearing. Nutrition enrichment mainly focuses only on fatty acids, especially *n*-3HUFA, EPA, and docosahexaenoic acid (DHA). Various methods of enrichment are used in hatcheries globally.

4.3 Methods for Nutritional Enrichment and Enrichment Diets

Nutritional enrichment in hatcheries was started as secondary culture by using *N. oculata* after the primary culture performed using baker's yeast. Previously, *N. oculata* was cultured in the outdoor ponds. Therefore, external factors such as weather and/or temperature influenced the production of *N. oculata*. This resulted in production instability of rotifers, and the use of baker's yeast was initiated to meet the nutritional requirements of rotifers. The nutritional state of baker's yeast was improved by Imada et al. (1979); they supplemented baker's yeast with fish oil or squid-liver oil that is rich in *n*-3 HUFAs. This method was called the indirect method, and some hatcheries still use this method. Watanabe et al. (1983) developed a direct method where lipids that include *n*-3 HUFAs such as fish oil or squid-liver oil were emulsified with egg yolk and seawater by using a juicer and introduced into baker's yeast before feeding to the rotifers. However, at present, emulsified oil or yeast is not used in many hatcheries. Recent commercial enrichment diet for rotifers, which is in a powdered or liquid form and contains *n*-3 HUFA, DHA, and/or EPA, needs to be mixed with water and then emulsified. The advantage of this method is that the contents of the enrichment diet can be varied. This allows the variation of *n*-3 HUFA contents according to the need of each fish species.

Some microalgae, including *n*-3 HUFA, DHA, and/or EPA, are used as enrichment diet. As mentioned before, *N. oculata* including rich EPA (Table 4.2) has been used as primary or enrichment diet. *N. oculata* is usually concentrated and then frozen or refrigerated. Recently *Schizochytrium* sp. was used as the enrichment diet (Barclay and Zeller 1997); it includes high amount of DHA. Although such enrichment diets cannot be varied to optimum contents for object fish species, they are suitable for fish species larvae that require large amounts of *n*-3 HUFA, EPA, or DHA, such as bluefin tuna (Seoka et al. 2007), *Seriola* spp. (Watanabe 1993), and Atlantic cod (Park et al. 2006). Moreover, feeding phytoplanktons that contain abundant nutritional contents, especially rich in *n*-3 HUFA, EPA, and DHA, is not only harmless for rotifers but also useful as an enrichment diet.

At present, *Chlorella vulgaris* is commonly used as feed for rotifers in many Japanese hatcheries. However, rotifers fed *C. vulgaris* require nutritional enrichment. Hayashi et al. (2001) succeeded in promoting exogenous uptake of DHA by *C. vulgaris*, and Yukino et al. (2004) reported the effect of enriched *C. vulgaris* on rotifers. *C. vulgaris* enriched with DHA and EPA contents is now commercialized.

4.4 Specific Nutrients for Enrichment

In the 1970–1980s, nutritional enrichment of rotifers mainly focused on *n*-3 HUFA. The growth and survival of larval fish were improved with the enrichment of *n*-3 HUFA, EPA, and DHA. However, the rotifers were also deficient in amino acid content. Hence, other kinds of nutrition such as vitamins (Miki et al. 1990) and phospholipid (Kanazawa et al. 1981, 1983) were used. Indeed, phospholipid, free amino acids, and minerals were effective in improving the health of young or adult fish (Chen et al. 2004; Nguyen et al. 2008; Penglase et al. 2011).

4.4.1 Fatty Acids

Most finfishes require *n*-3 fatty acids more than *n*-6 ones. EPA and DHA belong to *n*-3 fatty acids and are essential for animals. EPA is known as the precursor of physiological active substances, such as prostaglandin and thromboxane, and is essential in various physiological functions in fishes. DHA is included in the neural system and plays an important role in the development and function expression of fishes. Watanabe (1993) suggested that DHA is superior to EPA as essential fatty acid. On the other hand, EPA is an important source of eicosanoids though it is competitive as the source to arachidonic acid (ARA, 20:4 *n*-6; Sargent et al. 1999). Although it has been reported that EPA's role is inferior or equal to DHA and ARA, the interaction of EPA with DHA and EPA has become important.

Freshwater fishes can produce EPA from C18:3 *n*-3 but marine fishes cannot. Therefore, EPA and DHA are essential for marine fishes. Marine finfish larvae also require high amount of EPA and DHA. Specifically, the ratio of DHA rapidly decreases in larval body content of marine finfish larvae until 10 days post hatching (Zheng et al. 1995). Many studies showed that deficiency in DHA resulted to mortalities in larvae, abnormal behavior, and reduction of stress tolerance. Therefore, marine finfish larvae have to take these fatty acids exogenously.

n-3 HUFA, as well as EPA and DHA, have been used to enrich rotifers, although they naturally contain some amount of *n*-3 HUFAs. The enrichment of rotifers with *n*-3 HUFAs should be sufficient to meet the requirements of fish larvae (Izquierdo 1996).Since the required quantity of *n*-3 HUFA, EPA, or DHA varies across different fish species, the amount used for enrichment depend on the targeted species. Hence, if the supply of essential fatty acids is inadequate, then fish larvae will likely to show symptoms of such deficiencies (Izquierdo 1996). Nevertheless, some species such as mud crab larvae were found to develop abnormalities because of excess DHA supplementation (Suprayudi et al. 2002). This has not been reported in any other larval species. Further, not only the quantitative enrichment of fatty acids but also the composition of enrichment should be considered. The optimum DHA/EPA ratio for different fish species has been determined (Rodríguez et al. 1997; Copeman et al. 2002; Kotani et al. 2013). According to Rodríguez et al. (1997), 2.5/1 of ratio is optimum for gilthead sea bream larvae, whereas 1-2/1 is suitable for red sea bream (Kotani et al. 2013). For the larvae of some fish species that require higher level of DHA, the optimum ratio is also high. Such fishes include yellowtail flounder (Copeman et al. 2002) and Atlantic cod (Park et al. 2006). Although the DHA/EPA ratio in rotifers reflects that of the enrichment diet (Kotani et al. 2013), the fatty acid composition of rotifers does not completely reflect that of the enrichment diet. For example, the content of ARA increases after enrichment without inclusion in the enrichment diet (Kotani et al. 2013). The fatty acid composition of rotifers is influenced by the nutrition of microalgae used as their feeds. Although *C. vulgaris* does not include ARA (Maruyama et al. 1997), it is already contained in rotifers before enrichment. This suggests that rotifers have physiological systems that allow them to synthesize ARA. At present, information regarding the role of any bacteria in this phenomenon or that of internal metabolite is not yet known. Rotifers can synthesize fatty acids such as ARA as well as DHA (Lubzens et al. 1985). DHA, EPA, and ARA are known to be important components for fish larvae. Therefore, some studies have attempted to determine their composition ratio (DHA:EPA:ARA) (Sargent et al. 1999; Bell and Sargent 2003; Koven et al. 2003). Effective and appropriate nutritional enrichment of rotifers requires that further studies should determine the internal synthesis of fatty acids and the fatty acid components.

4.4.2 Phospholipid

The major lipid constituent of fish egg yolk or yolk-sac larvae is phospholipids (Sargent et al. 1999). If this is defined as the ideal lipid component of larval diet, phospholipids have to be included in rotifers as live feed. Kanazawa et al. (1981, 1983) pointed out the necessity of phospholipids for ayu, *Plecoglossus altivelis* larvae, by including phospholipids in the micro-artificial diet. Phospholipids mainly constitute the biomembrane, and therefore larvae developing rapidly request them (Hamre et al. 2013). It is well known that DHA is a component of the cell membrane of the neural system, and phospholipids including ARA and EPA are known to be source of eicosanoid. Dietary phospholipids, specifically phosphatidylcholine, have beneficial effects on larval growth and survival. Dietary phospholipids are necessary because they are known to enhance the absorption of ingested fats by acting as temporary emulsifier (Izquierdo and Koven 2011).

Generally, phospholipids are incorporated into rotifers, in the form of emulsified oil, commercial enrichment diet, or microalgal food (Fernández-Reiriz et al. 1993; Rainuzzo et al. 1994a; Fernández-Reiriz and Labarta 1996). However, it is known that enrichment with emulsified oil is not always effective (Rainuzzo et al. 1994b). Considering the lipid class of phospholipid, some studies reported that phosphatidylcholine gives positive effects for the larval rearing performance. On the other hand, some studies showed that the fatty acid composition in phospholipids should be taken into consideration and DHA should be enriched to phospholipid (Bell et al. 2003; Tocher et al. 2008). The amount of DHA to be enrich in phospholipid depends on the method of enrichment in rotifers as mentioned in 5-2, and it is necessary to improve that method of enrichment if the DHA content is required to be increase.

4.4.3 Vitamins

Many studies have shown that vitamins are necessary for fish larvae. Thus, vitamins such as fat (vitamins A and E)- and water-soluble (vitamin C) ones have been used to enrich rotifers, and the effect of such supplementations has been reported (Miki et al. 1990; Zheng et al. 1997; Hamre et al. 2008a, b). At present, vitamin A is considered important for larval health and quality, and hence, enrichment of vitamin A is mainly presented in this chapter.

Around the 1970s, albinism of hatchery-reared Japanese flounder was frequently observed. Miki et al. (1990) found that feeding rotifers enriched with vitamin A to larvae could reduce albinism. However, excess dietary vitamin A caused tail deformities (Miki et al. 1990). Such deformity was also noted in gilthead sea bream (Fernández et al. 2008), Atlantic halibut (Lewis-McCrea and Lall 2010), and European sea bass (Mazurais et al. 2009). Therefore, rotifers have to include suitable amount of vitamin A to avoid any abnormalities. Fushimi et al. (2005) recommended 212 IU/g of vitamin A in rotifers.

4.4.4 Free Amino Acids

Enrichment with protein has not yet been conducted. However, free amino acids are required for good fish larval nutrition. One of the most important free amino acids that can be used for rotifer enrichment is taurine. Taurine is a sulfur amino acid and plays an important role in organism tissue. It is an essential nutrient for red sea bream and Japanese flounder larvae (Takeuchi et al. 2001) since it improves their growth and survival (Chen et al. 2004, 2005). However, the amounts of taurine in rotifers are sufficient to ensure effective larval growth and health. Takahashi et al. (2005) established a rotifer enrichment method that can produce over 1000 mg/100 g dry weight of taurine.

4.4.5 Minerals

Minerals are responsible for skeletal formation, colloidal system, acid-base balance regulation, as well as hormone and enzyme regulation (Watanabe et al. 1997; Satoh 2003). Zinc and manganese deficiencies in fish cause skeletal deformities (Satoh et al. 1983; Watanabe et al. 1997; Nguyen et al. 2008), and selenium deficiency causes ataxia (Bell et al. 1986). Rotifers that are not enriched contain little amounts of minerals than that found in copepods (Hamre et al. 2008a). Matsumoto et al. (2009) cultured *C. vulgaris* in a medium that included zinc and then fed these *C. vulgaris* to rotifers. Penglase et al. (2011) enriched rotifers with selenium by using a commercialized yeast strain supplemented with selenium. Although Hamre et al. (2008b) enhanced rotifers with selenium and iodine by using a commercial enhancement diet containing sodium selenite and sodium iodide, no study has performed enrichment by using a mixture of important minerals such as iodine, manganese, selenium, and zinc.

4.5 Rotifer Quality and Enrichment Effect

4.5.1 Influence of Primary Culture

Producing healthy finfish larvae requires the stable supply of good quality and quantity of live food. Since the 1980s, rotifers have been used as the main live feed (Hino 1994). This is because mass culture and nutritional enrichment methods of rotifers had been established, and stable supply of rotifers having adequate quality was fulfilled by these strategies. However, rotifer productivity is known to fluctuate depending on the culture method and their population growth phase. Their nutritional value also fluctuates depending on the enrichment method. Therefore, methods need to be developed to stabilize these aspects.

Batch culture has already been recognized as an unstable method for rotifer production (Kotani et al. 2009). Kotani et al. (2009) reported that the enrichment effect, especially that of DHA and EPA contents, differed among the population growth phases in batch cultures. Rotifers in logarithmic phase had the highest content of these compounds. However, Tomoda et al. (2004, 2005) did not find the differences of nutritional enrichment effect among population growth phases, but they reported that feeding rotifers after the stationary phase in batch culture adversely affected the growth of red sea bream and Japanese flounder larvae.

Generally, when rotifers are cultured using the batch method, they are harvested in the late logarithmic or stationary phase and then enriched nutritionally. That is, during the harvest in batch culture, the most effective phase for enrichment is missed. Even if the rotifers are harvested during the logarithmic phase, the population number in the logarithmic phase cannot reach sufficient amount to feed larvae and to start a new culture. However, a continuous culture (Kuwada 2000) theoretically maintains the population growth at the logarithmic phase; therefore, the nutritional enrichment effect in rotifers cultured using the continuous culture method can be higher (Kotani et al. 2009).

Further, concentrations of free ammonia and dissolved oxygen in the culture medium affect the population growth and physiological activity of rotifers in batch culture (Koiso and Hino 2002; Kotani et al. 2009). During the progression of batch culture, the concentration of free ammonia increases and that of dissolved oxygen decreases (Koiso and Hino 2002; Kotani et al. 2009). Because the concentrations of free ammonia and dissolved oxygen are maintained relatively high during continuous culture, the DHA and EPA contents of enriched rotifers cultured using the continuous method are also high (Kotani et al. 2009). The influence of free ammonia on rotifer activity might be decreased when the concentration of dissolved oxygen is high. Therefore, obtaining efficient nutritional enrichment effect requires that the concentration of dissolved oxygen should be maintained high in the primary culture medium regardless of the culture method used.

4.5.2 Arrangement of Enrichment Method

Rotifers can be enriched by feeding them commercial enrichment diets. Each enrichment diet has recommendations regarding three directions for use, including quantities and shelf life. The recommended directions are possibly effective for providing optimum nutritional value for larvae, and many previous studies have followed these directions. Thus, obtaining rotifers with more nutritional value is possible by using higher doses of enrichment diets and for longer duration. Higher doses of enrichment diet can increase the DHA and EPA contents (Kotani et al. 2010), as well as enhance vitamin A content. Since the excess of vitamin A can cause skeletal deformities, using higher doses of enrichment diet should be avoided. On the other hand, longer duration is also effective to enrich DHA and/or EPA in phospholipid or polar lipid (Li and Olsen 2015; Kotani et al. 2017). This is true in the case of algae-based enrichment diet. Kotani et al. (2010) reported that rotifers cannot effectively absorb more nutrients beyond a certain limit with longer enrichment period using commercial enrichment diet or combination of some materials, e.g., fish oil and soybean protein. Li and Olsen (2015) conducted a 24-h enrichment period and Kotani et al. (2017) for 3 days. Both studies showed that enrichment is more effective with over 24-h enrichment using algae-based diet.

Each commercial enrichment diet is not developed by mixing it with other diets. However, some enrichment diet produces complex effects; they can be considered to contain more than one constituent. Matsumoto et al. (2009) enriched rotifers with a commercial enrichment diet and *C. vulgaris* with Zn. Therefore, the rotifers did not contain sufficient amounts of DHA and EPA, but were enriched only with Zn. This suggests that rotifers preferentially ingested *Chlorella* over the emulsified enrichment diet (Matsumoto et al. 2009). This is because the mastax of rotifers can actively ingest live foods but not abiotic materials (Funamoto and Hirayama 1982).

Therefore, enrichment diets with varying treatments and including many constituents cannot be used. Each enrichment diet should be used as per the method recommended by the manufacturer. Although enrichment diets should not be mixed, their mixing usage is inevitable, especially when one enrichment diet is limited in some nutrients. For example, enrichments for taurine and zinc or manganese can be combined. Nevertheless, developing a nutritional enrichment diet that contains all the required nutrients in well-balanced form is important.

4.5.3 Difference in Enrichment Among Rotifer Species

B. plicatilis sp. complex has various characteristics and differ among strains or species. In the past, rotifer populations were grouped based on lorica size; those with large lorica were classified as "L type" and small as "S type." These were described as different species (Segers 1995). Subsequently, rotifer *Brachionus* populations were divided into 15 species (Mills et al. 2016). The physiological characteristics of all these 15 rotifer species have not been found, but some of them have been clarified already. For example, L-type rotifers show high fecundity at water temperature range of 20–25 °C, while S type at 25–30 °C (Hirayama and Rumengann 1993). This reflects the physiological difference among species. Such difference is inferred as a result in the difference in the nutritional enrichment performance among rotifer species. Especially the absorbance of nutritional contents in enrichment diet is influenced by the physiological activities of rotifers. Moreover, Yúfera et al. (1997) reported that the energy content on a dry weight basis differs between (S type) *B. rotundiformis* and L type (*B. plicatilis*). Li et al. (2015) reported that lipid ration is different between *Brachionus* Cayman and *Brachionus* Nevada, and such difference can be based on the size difference among these species. Moreover, although L-type rotifers enriched or treated for longer time (>24 h) with the commercial *C. vulgaris* product including DHA can increase the DHA content in polar lipid, S-type rotifers cannot show such increment (Kotani et al. 2017). Few similar reports are available on the differences in nutritional value among rotifer species as mentioned above. After it is recognized that rotifer populations can be divided into 15 species, the differences of nutritional enrichment performance among rotifer species should be the important information for the manufacturers of enrichment diet, and every hatcheries should know about these differences. Further studies are expected to be related in this topic.

4.5.4 Nutritional Changes of Rotifers in the Rearing Water

During larviculture, nutritionally enriched rotifers are fed to fish larvae two or three times a day. From one feeding to the next feeding, the nutritional contents of the enriched rotifers leached or degraded into the rearing water. It is known that DHA and/or EPA in rotifers degrade 6–12 h after the enrichment treatment (Dan and Koiso 2008). To avoid nutritional deterioration or leaching, microalgae such as *N. oculata* or *C. vulgaris* which serves as food for rotifers should also be enriched with *n*-3 HUFAs (Dan and Koiso 2008). Not only fatty acid contents but also protein and carbon contents also decreased in water when >12 h has passed since the inoculation of rotifers just after nutritional enrichment (Øie et al. 1997). It is known that the addition of algae *Isochrysis galbana* can maintain the protein and carbon contents of rotifers in the rearing water. This method is known as "green water," and studies showed that this method is effective in improving the survival and the growth of cultured larvae (Øie et al. 1997; Faulk and Holt 2005; van der Meeren et al. 2007). These studies suggested that rotifers consumed the enriched nutrition by their metabolism, and this consumption resulted in the deficiency of nutrition. On the other hand, during starvation, there is a slight decrease in phospholipid in rotifers, but triacylglycerol content considerably decreased (Rainuzzo et al. 1994b). As mentioned in 4–1 and 4–2, the fatty acid contents in phospholipids are important to improve the larval rearing performance. Furthermore, keeping the enriched rotifer in "green water" to maintain its nutritional value is also important.

4.6 Prospects

The nutritional enrichment of rotifers has been performed using the direct method and by feeding phytoplankton or other organisms that have high nutritional value to the rotifers. The chemical contents of each animal are influenced by the nutritional value of its feeds. Therefore, the usage of phytoplankton or organisms as feed is convenient for the nutritional enrichment of rotifers. However, the nutrient content of phytoplankton might not always be adequate for the target species of finfish larviculture that have diversified nutritional needs. Further, since each phytoplankton includes different nutritional contents at various population growth phase (Okauchi et al. 1990), assuring the qualitative stability of the nutritional value of rotifers after nutritional enrichment, especially of those reared using biomass culture in a hatchery, is difficult. The direct method can be used to obtain nutritional enrichment diet produced from various materials and by using specific chemical components. Therefore, rotifers enriched by using the direct method can be used for various fish species larvae. Hence, nutritional enrichment using the direct method might become more common in the future.

Present enrichment diets can ensure that rotifers acquire enough amount of nutrition, especially DHA and EPA. However, enrichment with phospholipids or polar lipids seems insufficient. Although the present commercial enrichment diets can be sufficiently enriched in DHA and EPA (nonpolar lipids), copepods or larval fishes just after hatch contain a certain amount of DHA and EPA in the phospholipids or polar lipids, and rotifers enriched by using commercial enrichment diets cannot contain DHA and EPA in polar lipids in amounts as large as those found in copepods or larval fishes (Kotani, unpublished; Table 4.3). Regarding copepods or larval fishes as the model of feeds in finfish larviculture, rotifers should be enriched with DHA and EPA into polar lipids while ensuring balanced nutrition for rotifers. The balanced nutrition should include fatty acids, amino acids, vitamins, and minerals. Thus, a capsule of completely balanced enrichment diet should be developed. Further, studies should investigate whether rotifers can acquire all the nutrients from the balanced nutrition.

Table 4.3 Percentages of total nonpolar and polar lipids and the fatty acid composition of each type of lipid in the live feeds of finfish

Type of feeds (finfish larvae)	Copepod from the Arctic Circle		Enriched *B. plicatilis*		Blue emperor *Lethrinus nebulosus* larvae		Striped beakfish *Oplegnathus fasciatus* larvae	
Type of lipid (% in total lipid)	Nonpolar	Polar	Nonpolar	Polar	Nonpolar	Polar	Nonpolar	Polar
	45.68	54.32	59.49	40.51	65.44	34.56	66.95	33.05
Fatty acid composition (%)								
14:0	4.29	5.94	2.36	1.38	5.21	1.96	3.27	1.32
16:0	19.67	21.57	12.51	17.74	18.91	23.98	17.57	22.09
16:1	9.08	4.42	0.48	0.66	6.14	1.40	9.05	1.61
16:2	2.36	0.69	0.71	0.60	0.47	0.19	0.34	0.20
16:3	1.38	0.77	0.02	0.82	0.12	0.00	0.05	0.03
18:0	3.45	4.38	2.80	3.90	7.53	7.24	3.58	7.32
18:1 n-9	9.56	3.52	10.39	6.23	18.84	9.23	19.21	7.55
18:1 n-7	2.34	1.74	1.71	1.63	3.72	2.06	4.96	2.50
18:2 n-6	9.11	1.76	6.25	12.71	2.55	1.59	2.73	1.23
18:3 n-3	0.85	0.48	1.35	2.60	0.67	0.20	0.57	0.12
18:4 n-3	3.43	0.78	–	–	1.13	0.23	0.66	0.10
20:4 n-6	0.93	0.99	1.86	2.25	1.11	3.36	1.34	4.29
20:4 n-3	0.88	11.67	1.61	2.70	0.68	0.32	1.19	0.45
20:5 n-3	5.38	0.60	7.88	6.47	4.86	6.96	5.22	7.44
22:0	0.53	0.22	–	–	0.13	0.22	–	–
22:5 n-6	0.41	0.43	6.39	2.19	0.57	0.95	0.56	0.65
22:5 n-3	1.10	0.91	3.58	5.46	1.85	1.38	3.41	3.04
22:6 n-3	4.80	19.52	27.06	12.53	18.86	33.95	17.90	33.87

References

Barclay W, Zeller S (1997) Nutritional enhancement of n-3 and n-6 fatty acids in rotifers and *Artemia* nauplii by feeding spray-dried *Schizochytrium* sp. J World Aquacult Soc 27:314–322

Bell JG, Sargent JR (2003) Arachidonic acid in aquaculture feeds: current status and future opportunities. Aquaculture 218:491–499

Bell JG, Pirie BJS, Adron JW, Cowey CB (1986) Some effects of selenium deficiency on glutathione peroxidase (*EC* 1.11.1.9) activity and tissue pathology in rainbow trout (*Salmo gairdneri*). Br J Nutr 55:305–311

Bell JG, McEvoy LA, Estevez A, Shields RJ, Sargent JR (2003) Optimising lipid nutrition in first-feeding flatfish larvae. Aquaculture 227:211–220

Chen J-N, Takeuchi T, Takahashi T, Tomoda T, Koiso M, Kuwada H (2004) Effect of rotifers enriched with taurine on growth and survival activity of red sea bream *Pagrus major* larvae. Nippon Suisan Gakkaishi 70:542–547. (in Japanese with English abstract)

Chen J-N, Takeuchi T, Takahashi T, Tomoda T, Koiso M, Kuwada H (2005) Effect of rotifers enriched with taurine on growth in larvae Japanese flounder *Paralichthys olivaceus*. Nippon Suisan Gakkaishi 71:342–347. (in Japanese with English abstract)

Copeman LA, Parrish CC, Brown JA, Harel M (2002) Effects of docosahexaenoic, eicosapentaenoic, and arachidonic acids on the early growth, survival, lipid composition and pigmentation of yellowtail flounder (*Limanda ferruginea*): a live food enrichment experiment. Aquaculture 210:285–304

Dabrowski K (1984) The feeding of fish larvae: present and perspective. Repord Nutr Dévelop 24:807–833

Dan S, Koiso M (2008) Effect of microalgal addition on stability of n-3HUFA contents in enriched rotifer *Brachionus plicatilis* in large tank for seed production. Aquacult Sci 56:603–604. (in Japanese with English abstract)

Faulk CK, Holt GJ (2005) Advances in rearing cobia *Rachycentron canadum* larvae in recirculating aquaculture systems: live prey enrichment and greenwater culture. Aquaculture 249:231–243

Fernández I, Hontoria F, Ortiz-Delgado JB, Kotzamanis Y, Estévez A, Zambonino-Infante JL, Gisbert E (2008) Larval performance and skeletal deformities in farmed gilthead sea bream (*Sparus aurata*) fed with graded levels of vitamin A enriched rotifers (*Brachionus plicatilis*). Aquaculture 286:102–115

Fernández-Reiriz MJ, Labarta U (1996) Lipid classes and fatty acid composition of rotifers (*Brachionus plicatilis*) fed two algal diets. Hydrobiologia 330:73–79

Fernández-Reiriz MJ, Labarta U, Ferreiro MJ (1993) Effects of commercial enrichment diets on the nutritional value of the rotifer (*Brachionus plicatilis*). Aquaculture 112:195–206

Fukusho K (1977) Nutritional effects of the rotifer, *Brachionus plicatilis*, raised by baking yeast on larval fish of *Oplegnathus fasciatus*, by enrichment with *Chlorella* sp. before feeding. Bull Nagasaki Prefectural Inst Fish 3:152–154. (in Japanese)

Funamoto H, Hirayama K (1982) Food selectivity of the rotifer *Brachionus plicatilis* expressed by using the frequency of mastax movement as an index. Suisanzoshoku 29:246–250

Fushimi H, Kotani T, Suzuki H, Hayashi M (2005) Life stage and morphogenesis of finfish larvae, with special reference to improvement of larval health. Eur Aquacult Soc Spec Publ 36:157–159

Hamre K, Srivastava A, Rønnestad I, Mangor-Jensen A, Stoss J (2008a) Several micronutrients in the rotifer *Brachionus* Sp. may not fulfil the nutritional requirements of marine fish larvae. Aquac Nutr 14:51–60

Hamre K, Mollan TA, Sæle Ø, Erstad B (2008b) Rotifers enriched with iodine and selenium increase survival in Atlantic cod (*Gadus morhua*) larvae. Aquaculture 284:190–195

Hamre K, Yúfera M, Rønnestad I, Boglione C, Conceição LEC, Izquierdo M (2013) Fish larval nutrition and feed formulation: knowledge gaps and bottlenecks for advances in larval rearing. Rev Aquac 5:S26–S58

Hayashi M, Yukino T, Maruyama I, Kido S, Kitaoka S (2001) Uptake and accumulation of exogenous docosahexaenoic acid by *Chlorella*. Biosci Biotechnol Biochem 65:202–204

Hino A (1994) Aquaculture – seed production. In: "Recent advances in fisheries science" (ed. by The publication committee in the Japanese Society of Fisheries Science), Koseisha Koseikaku, Tokyo, pp 124–131 (in Japanese)

Hirata H (1964) Cultivation of live feed in Yashima Station-1. Saibai-Gyogyo News 2:4. (in Japanese)

Hirata H, Mori Y (1967) Cultivation of rotifer *Brachionus plicatilis* with feeding baker's yeast. Saibai-Gyogyo 5:36–40. (in Japanese)

Hirayama K, Rumengann IM (1993) The fecundity patterns of S and L type rotifers of *Brachionus plicatilis*. Hydrobiologia 255-256:153–157

Hunter JR (1981) Feeding ecology and predation of marine fish larvae. In: Lasker R (ed) Marine fish larvae: morphology, ecology, and relation to fisheries. University of Washington Press, Seattle, pp 34–77

Imada O, Kageyama Y, Watanabe T, Kitajima C, Fujita S, Yone Y (1979) Development of a new yeast as a culture medium for living feeds used in the production of fish seed. Nippon Suisan Gakkaishi 45:955–959

Ito T (1960) On the culture of mixohaline rotifer *Brachionus plicatilis* O. F. Müller in the sea water. Rep Facul Fish Prefect Univ Mie 3:708–740. (in Japanese)

Izquierdo MS (1996) Essential fatty acid requirements of cultured marine fish larvae. Aquac Nutr 2:183–191

Izquierdo M, Koven W (2011) Lipids. In: Larval fish nutrition (ed by Holt GJ). Wiley-Blackwell, pp 47–81

Jaroszewska M, Dabrowski K (2011) Utilization of Yolk: transition from endogenous to exogenous nutrition in fish. In: Larval fish nutrition. Wiley-Blackwell, pp 183–218.

Kanazawa A, S-I T, Inamori S, Iwashita T, Nagao A (1981) Effects of phospholipids on growth, survival rate, and incidence of malformation in the larval ayu. Mem Fac Fish Kagoshima Univ 30:301–309

Kanazawa A, S-i T, Kobayashi T, Takae M, Iwashita T, Uehara R (1983) Necessity of dietary phospholipids for growth of the larval ayu. Mem Fac Fish Kagoshima Univ 32:115–120

Kitajima C, Koda T (1976) Lethal effects of the rotifer cultured with baking yeast on the larval red sea bream, *Pagrus major*, and the increase of survival rate using the rotifer recultured with *Chlorella* sp. Bull Nagasaki Prefect Inst Fish 2:113–116. (in Japanese)

Kitajima C, Fujita S, Owa F, Yone Y, Watanabe T (1979) Improvement of dietary value for red sea bream larvae of rotifers *Brachionus plicatilis* cultured with baker's yeast *Saccharomyces cerevisiae*. Nippon Suisan Gakkaishi 45:469–471. (in Japanese with English abstract)

Kjørsvik E, Galloway TF, Estevez A, Sæle Ø, Moren M (2011) Effects of larval nutrition on development. In: Larval fish nutrition. Wiley-Blackwell, pp 219–248.

Koiso M, Hino A (2002) Biological examinations on growth stagnation mechanism in the mass-producing batch culture of the rotifer, *Brachionus plicatilis*. Suisanzoshoku 50:197–204. (in Japanese with English abstract)

Kolkovski S (2001) Digestive enzymes in fish larvae and juveniles – implications and applications to formulated diets. Aquaculture 200:181–201

Kotani T, Genka T, Fushimi H, Hayashi M, Dierckens K, Sorgeloos P (2009) Effect of primary cultivation method on nutritional enrichment of euryhaline rotifer *Brachionus plicatilis*. Fish Sci 75:975–984

Kotani T, Genka T, Tanabe M, Miyashima A, Fushimi H, Hayashi M (2010) Effect of nutritional enrichment method on fatty acid contents of rotifer *Brachionus plicatilis*. J World Aquacult Soc 41:884–892

Kotani T, Fushimi H, Ohta Y, Miyashima A, Sudoh K, Hayashi M, Satoh N, Satoh S (2013) Effect of graded levels of dietary DHA included in rotifers *Brachionus plicatilis* on larviculture performance of red sea bream *Pagrus major*. Aquacult Sci 61:321–330

Kotani T, Haraguchi T, Yamazaki Y, Doi T, Matsui H, Yokoyama S, Ishikawa M, Koshio S (2017) Effect of the time period of nutritional enrichment on the fatty acid composition of commonly used rotifers *Brachionus plicatilis* sp. complex and larviculture performance of red sea bream *Pagrus major*. Aquacult Sci 65:133–144

Koven W, van AnholtR LS, Atia IB, Nixon O, Ron B, Tandler A (2003) The effect of dietary arachidonic acid on growth, survival, and cortisol levels in different-age gilthead seabream larvae (*Sparus auratus*) exposed to handling or daily salinity change. Aquaculture 228:307–320

Kuwada H (2000) Extensive continuous culture – recent developed continuous culture. In: Japan Sea Farming Association (ed) Guidebook of cultivation of marine rotifer Brachionus. Japan Sea Farming Association, Tokyo, pp 92–107. (in Japanese)

Lewis-McCrea LM, Lall SP (2010) Effects of phosphorus and vitamin C deficiency, vitamin A toxicity, and lipid peroxidation on skeletal abnormalities in Atlantic halibut (*Hippoglossus hippoglossus*). J Appl Ichthyol 26:334–343

Li K, Olsen Y (2015) Effect of enrichment time and dietary DHA and non-highly unsaturated fatty acid composition on the efficiency of DHA enrichment in phospholipid of rotifer (*Brachionus* Cayman). Aquaculture 446:310–317

Li K, Kjørsvik E, Bergvik M, Olsen Y (2015) Manipulation of the fatty acid composition of phosphatidylcholine and phosphatidylethanolamine in rotifers *Brachionus* Nevada and *Brachionus* Cayman. Aquac Nutr 21:85–97

Lubzens E, Marko A, Tietz A (1985) De novo synthesis of fatty acids in the rotifer, *Brachionus plicatilis*. Aquaculture 47:27–37

Maruyama I, Nakao T, Shigeno I, Ando Y, Hirayama K (1997) Application of unicellular *Chlorella vulgaris* for the mass-culture of marine rotifer *Brachionus*. Hydrobiologia 358:133–138

Matsumoto S, Satoh S, Kotani T, Fushimi H (2009) Examination of a practical method for zinc enrichment of euryhaline rotifers(*Brachionus plicatilis*). Aquaculture 286:113–120

Mazurais D, Glynatsi N, Darias MJ, Christodoulopoulou S, Cahu CL, Zambonino-Infante J-L, Koumoundouros G (2009) Optimal levels of dietary vitamin A for reduced deformity incidence during development of European sea bass larvae (*Dicentrarchus labrax*) depend on malformation type. Aquaculture 294:262–270

Miki N, Taniguchi T, Hamakawa H, Yamada Y, Sakurai N (1990) Reduction of albinism in hatchery-reared flounder "Hirame", *Paralichthys olivaceus* by feeding on rotifer enriched with vitamin-A. Suisanzoshoku 38:147–155. (in Japanese with English abstract)

Mills S, Alcántara-Rodríguez AJ, Ciros-Pérez J, Gómez A, Hagiwara A, Galindo KH, Jersabek CD, Malekzadeh-Viayeh R, Leasi F, Lee J-S, Mark Welch DB, Papakostas S, Riss S, Segers H, Serra M, Shiel R, Smolak R, Snell TW, Stelzer C-P, Tang CQ, Wallace RL, Fontaneto D, Walsh EJ (2016) Fifteen species in one: deciphering the *Brachionus plicatilis* species complex (Rotifera, Monogononta) through DNA taxonomy. Hydrobiologia:1–20

Nguyen VT, Satoh S, Haga Y, Fushimi H, Kotani T (2008) Effect of zinc and manganese supplementation in *Artemia* on growth and vertebral deformity in red sea bream (*Pagrus major*) larvae. Aquaculture 285:184–192

Okauchi M, Zhou W-J, Zou W-H, Fukusho K, Kanazawa A (1990) Difference in nutritive value of a microalga *Nannochloropsis oculata* at various growth phases. Nippon Suisan Gakkaishi 56:1293–1298. (in Japanese with English abstract)

Øie G, Makridis P, Reitan KI, Olsen Y (1997) Protein and carbon utilization of rotifers (*Brachionus plicatilis*) in first feeding of turbot larvae (*Scophthalmus maximus* L.) Aquaculture 153:103–122

Park H-G, Puvanendran V, Kellett A, Parrish CC, Brown JA (2006) Effect of enriched rotifers on growth, survival, and composition of larval Atlantic cod (*Gadus morhua*). ICES J Mar Sci 63:285–295

Penglase S, Hamre K, Sweetman JW, Nordgreen A (2011) A new method to increase and maintain the concentration of selenium in rotifers (*Brachionus* spp.) Aquaculture 315:144–153

Rainuzzo JR, Reitan KI, Jørgensen L, Olsen Y (1994a) Lipid composition in turbot larvae fed live feed cultured by emulsions of different lipid classes. Comp Biochem Physiol Part A 107:699–710

Rainuzzo JR, Reitan KI, Olsen Y (1994b) Effect of short- and long-term lipid enrichment on total lipids, lipid class and fatty acid composition in rotifers. Aquacult Int 2:19–32

Rodríguez C, Pérez JA, Díaz M, Izquierdo MS, Fernández-Palacios H, Lorenzo A (1997) Influence of the EPA/DHA ratio in rotifers on gilthead seabream (*Sparus aurata*) larval development. Aquaculture 150:77–89

Sargent J, McEvoy L, Estevez A, Bell G, Bell M, Henderson J, Tocher D (1999) Lipid nutrition of marine fish during early development: current status and future directions. Aquaculture 179:217–229

Satoh S (2003) Minerals. In: Makagawa H, Sato M (eds) Micronutrients and health of cultured fish. Koseisha-Koseikaku, Tokyo, pp 22–30. (in Japanese)

Satoh S, Takeuchi T, Narabe Y, Watanabe T (1983) Effects of deletion of several trace elements from fish meal diets on growth and mineral composition of rainbow trout fingerlings. Nippon Suisan Gakkaishi 49:1909–1916. (in Japanese with English abstract)

Segers H (1995) Nomenclatural consequences of some recent studies on *Brachionus plicatilis* (Rotifera, Brachionidae). Hydrobiologia 313–314:121–122

Seoka M, Kurata M, Kumai H (2007) Effect of docosahexaenoic acid enrichment in *Artemia* on growth of Pacific bluefin tuna *Thunnus orientalis* larvae. Aquaculture 270:193–199

Sumida S, Owaki M, Urata K (1974) Mass mortality in the progress during the larval rearing of red sea bream and Japanese parrotfish. Oper Rep Kumamoto Prefect Fish Exp Station 1973:373–382

Suprayudi MA, Takeuchi T, Hamasaki K, Hirokawa J (2002) The effect of n-3HUFA content in rotifers on the development and survival of mud crab, *Scylla serrata*, larvae. Suisanzoshoku 50:205–212

Sutton TM, Bowen SH (1994) Significance of organic detritus in the diet of larval lampreys in the Great Lakes Basin. Can J Fish Aquat Sci 51:2380–2387

Takahashi T, Amano T, Takeuchi T (2005) Establishment of direct enrichment method of taurine to rotifer. Aquacult Sci 53:121–126. (in Japanese with English abstract)

Takeuchi T, Park G-S, Seikai T, Yokoyama M (2001) Taurine content in Japanese flounder *Paralichthys olivaceus* T. & S. And red sea bream *Pagrus major* T. & S. During the period of seed production. Aquacult Res 32(Suppl. 1):244–248

Tocher DR, Bendiksen EÅ, Campbell PJ, Bell JG (2008) The role of phospholipids in nutrition and metabolism of teleost fish. Aquaculture 280:21–34

Tomoda T, Koiso M, Kuwada H, Chen JN, Takeuchi T (2004) Dietary value of marine rotifer *Brachionus plicatilis* in different population growth stages for larval red seabream *Pagrus major*. Nippon Suisan Gakkaishi 70:573–582. (in Japanese with English abstract)

Tomoda T, Koiso M, Kuwada H, Chen JN, Takeuchi T (2005) Dietary value of marine rotifer *Brachionus plicatilis* at different population growth stages for larval Japanese flounder *Paralichthys olivaceus*. Nippon Suisan Gakkaishi 71:555–562. (in Japanese with English abstract)

van der Meeren T, Mangor-Jensen A, Pickova J (2007) The effect of green water and light intensity on survival, growth and lipid composition in Atlantic cod (*Gadus morhua*) during intensive larval rearing. Aquaculture 265:206–217

Watanabe T (1993) Importance of docosahexaenoic acid in marine larval fish. J World Aquacult Soc 24:152–161

Watanabe T, Kiron V (1994) Prospects in larval fish dietetics. Aquaculture 124:223–251

Watanabe T, Kitajima C, Arakawa T, Fukusho K, Fujita S (1978a) Nutritional quality of rotifer *Brachionus plicatilis* as a living feed from the viewpoint of essential fatty acids for fish. Nippon Suisan Gakkaishi 44:1109–1114. (in Japanese with English abstract)

Watanabe T, Arakawa T, Kitajima C, Fukusho K, Fujita S (1978b) Proximate and mineral composition of living feeds used in seed production of fish. Nippon Suisan Gakkaishi 44:979–984

Watanabe T, Tamiya T, Oka A, Hirata M, Kitajima C, Fujita S (1983) Improvement of dietary value of live foods for fish larvae by feeding them on ω3 highly unsaturated fatty acids and fat-soluble vitamins. Nippon Suisan Gakkaishi 49:471–479

Watanabe T, Kiron V, Satoh S (1997) Trace minerals in fish nutrition. Aquaculture 151:185–207

Yako LA, Dettmers JM, Stein RA (1996) Feeding preferences of omnivorous gizzard Shad as influenced by fish size and zooplankton density. Trans Am Fish Soc 125:753–759

Yúfera M, Darias MJ (2007) The onset of exogenous feeding in marine fish larvae. Aquaculture 268:53–63

Yúfera M, Parra G, Pascual E (1997) Energy content of rotifers (*Brachionus plicatilis* and *Brachionus rotundiformis*) in relation to temperature. Hydrobiologia 358:83–87

Yukino T, Hayashi M, Yoshimatsu T, Maruyama I, Murata H (2004) Nutritional enrichment and cultivation of rotifers by feeding of docosahexaenoic acid-enriched *Chlorella vulgaris* K-22. Suisanzoshoku 52:381–386

Zheng F, Takeuchi T, Yoseda K, Hirokawa J, Watanabe T (1995) Changes in fatty acid composition of cod larvae in mass production. Nippon Suisan Gakkaishi 61:756–761. (in Japanese with English abstract)

Zheng F, Takeuchi T, Kobayashi M, Hirokawa J, Watanabe T (1997) A study of vitamin E absorption in rotifer and cod larvae and the effect of DHA content in rotifer for cod larvae. Nippon Suisan Gakkaishi 63:77–84. (in Japanese with English abstract)

Chapter 5
Application of Rotifers for Larval Rearing of Marine Fishes Cultivated Under Various Conditions

Yoshitaka Sakakura

Abstract Two aspects of rotifer for larviculture were reviewed: (1) the recent findings that the condition (viability) of rotifers affects the health of marine fish larvae and (2) the quantitative approach of flow field in the larviculture tanks determining the relationships between water flow and distribution of rearing organisms that are paid less attention rather than culture methods of rotifers and nutritional enrichment of rotifers for larviculture. The growth phase of rotifers affects the quality of marine fish larvae. Not only survival and growth but also occurrence of malformation of fish larvae is significantly improved when they are fed with rotifers in exponential growth phase. Recent quantitative studies on the relationship between flow field in the larviculture tanks and distribution of rotifers and fish larvae revealed that manipulating the distribution and behavior of rotifers in a fish rearing tank by modifying water flow, food source, and light conditions is also important, in order to enhance rotifer viability and feeding incidence of marine fish larvae.

5.1 Introduction

In 1960, Itoh (1960) showed that the rotifer *Brachionus plicatilis* can be acclimatized to high salinity and can be mass cultured; since then, rotifers have been used as live feed for larviculture of marine fishes. Many mass culture methods for rotifers have been developed (see Chap. 3), and rotifers, especially *Brachionus plicatilis* species complex, have become an essential part of feeding during the larval stages of marine fishes and crustaceans (Lubzens 1987; Hagiwara et al. 2001; Conceição et al. 2010). During marine fish larviculture, successful initiation of the first feeding of rotifers to larvae is necessary in order to prevent early mortality, which is one of the most important aspects for ensuring effective production of the target species.

Y. Sakakura (✉)
Graduate School of Fisheries and Environmental Sciences, Nagasaki University, Nagasaki, Japan
e-mail: sakakura@nagasaki-u.ac.jp

A. Hagiwara, T. Yoshinaga (eds.), *Rotifers*, Fisheries Science Series,
DOI 10.1007/978-981-10-5635-2_5

Many studies have attempted to enhance the first feeding success of larvae by selecting the optimal prey size and/or prey species (Hagiwara et al. 2001, 2007). Although rotifers are ingested and digested by many marine fishes in their early life stages, nutritional manipulation (enrichment) is necessary for improving the quality of fish, that is, fish health (Tsukamoto et al. 1999). Highly unsaturated fatty acids (HUFAs) such as arachidonic acid (ARA), eicosapentaenoic acid (EPA), and docosahexaenoic acid (DHA) are known to be essential for marine fish larvae. Rotifers cultured using the conventional method do not have adequate contents of HUFAs for ensuring the health quality of larvae (Izquierdo et al. 1989; Watanabe et al. 1989). Therefore, numerous studies have been investigating methods to improve the nutritional quality of live feeds by nutritional enrichment of rotifers (described in Chap. 4; e.g., Bell et al. 2003; Hamre et al. 2013; Izquierdo 1996; Mæhre et al. 2013; Rainuzzo et al. 1997; Takeuchi 2001, 2014; Tocher 2010; Watanabe 1993).

Nutritional enrichment of rotifers for marine fish larvae, using rotifers as "microcapsule" to carry specific materials into fish larvae, has significantly improved the results of larviculture in terms of survival, development, and growth. After having recognized the significance of nutritional enrichment for marine fish larvae utilizing rotifers, many fish culturists in Japanese hatcheries often point out that the results of larviculture differ according to not only the nutritional enrichment of rotifers but also the quality and viability of rotifers themselves. They also claim that the distribution of fish larvae and live feeds including rotifers is different in the rearing tanks and that the management of rotifer density where fish larvae are distributed is an important factor for the success of initiating the active feeding of fish larvae. However, these two issues are based on the empirical assumption of fish culturists, and science-based information is very limited. Therefore, I will review in this chapter following two aspects that are paid less attention rather than culture methods of rotifers (Chap. 3) and nutritional enrichment of rotifers for larviculture (Chap. 4): (1) the recent findings that the condition (viability) of rotifers affects the health of marine fish larvae and (2) the quantitative approach of flow field in the larviculture tanks determining the relationships between water flow and distribution of rearing organisms.

5.2 The Effects of Rotifer Viability on the Quality of Marine Fish Larvae

In many Japanese hatcheries, rotifers are cultured by using the batch culture system or the semicontinuous system (see Chap. 3). In these culture systems, rotifers are harvested at one of the following three phases of population growth: (1) lag phase, (2) log phase (exponential phase), and (3) stationary phase. Harvested rotifers are then enriched and finally fed to fish larvae. Most researches on rotifer quality for fish larvae focused on how rotifers can be considered as transporters of enrichment substances. For instance, Kotani et al. (2009) reported that the enrichment efficiency was higher when the rotifers were cultured using continuous culture than those cultured using batch culture. However, few studies have focused on the conditions of rotifers that might affect the quality of fish larvae when used as live feed.

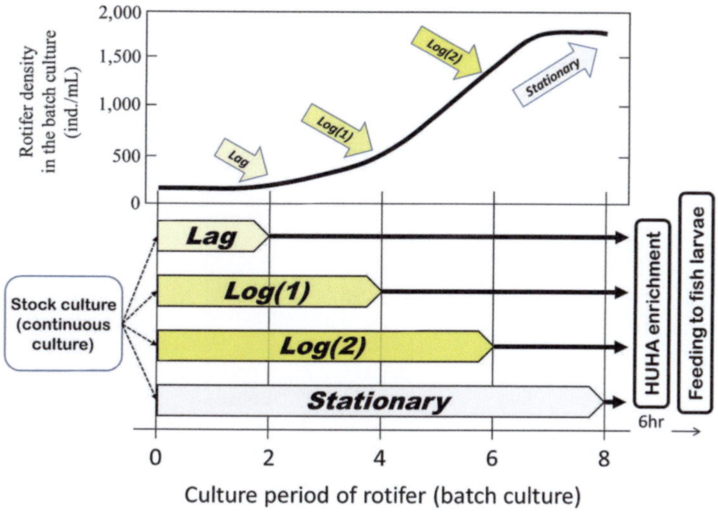

Fig. 5.1 Schematic drawings of preparation for rotifers at different population growth phase by Tomoda et al. (2004). Rotifers at different population growth phases were harvested by sliding the inoculation date from the stock culture of rotifers and enriched for fish larvae

Tomoda et al. (2004) firstly quantified the effects of population growth phase of rotifers on marine fish larviculture by using red sea bream *Pagrus major* larvae, which is one of the major aquaculture species in Japan. They hypothesized that viability of rotifers may be high at the exponential growth phase (log phase) and viability of rotifers will affect the performance of fish larvae. Then they prepared rotifers at different growth phases as follows (Fig. 5.1): L-type rotifer *Brachionus plicatilis* species complex was cultured using continuous culture as a stock. Then, rotifers were inoculated into a 500-L tank at a density of 100 ind./mL, and batch culture was performed by feeding *Chlorella vulgaris*. A total of nine batch culture tanks were prepared, and rotifers were inoculated with 1-day interval, and culture was performed for 8 days for each tank. The tank at 8-day culture was renewed and inoculated by the rotifers from the stock culture. With this rotation of batch culture, rotifers at different days after inoculation, thereby rotifers at different growth phases were continuously prepared. Rotifers were categorized into four growth phases: lag phase at day 2 (ca. 300 ind./mL), log (1) phase at day 4 (500 ind./mL), log (2) phase (1000 ind./mL), and stationary phase (1800 ind./mL) (Fig. 5.1). These rotifers at four different growth phases were then enriched with DHA-enriched *Chlorella vulgaris* for the feeding experiment to red sea bream larvae at 8 days after hatching (2.33 ± 0.05 mm in total length). Triplicate tanks containing 11,000 red seabream larvae were assigned for each growth phase of rotifers, and feeding experiment was conducted until 20 days after hatching. The proximate composition and fatty acid composition did not differ significantly among the rotifers at four growth phases or larvae fed with these rotifers. The survival rate (about >90%) was also the same among the feeding treatments. However, the growth of red sea bream larvae differed according to the growth phase of rotifers, and larvae fed with rotifers in the station-

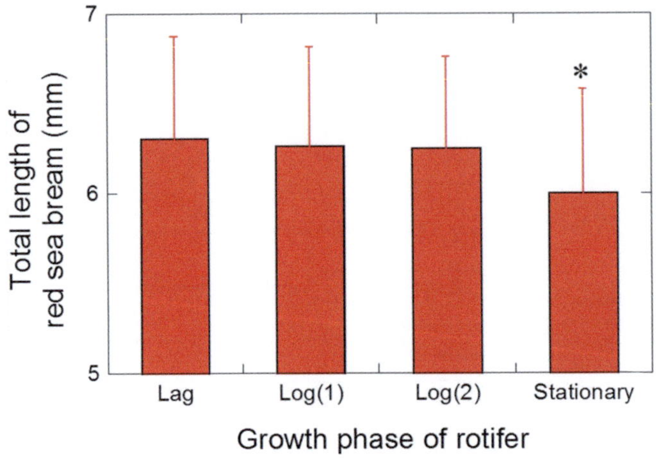

Fig. 5.2 Growth of red sea bream larvae at 20 days after hatching fed with rotifers from different growth phases (see Fig. 5.1). An *asterisk* indicates significant difference among rotifer growth phases ($n = 3$, Scheffe's F, $p < 0.05$) (Redrawn from Tomoda et al. 2004)

Fig. 5.3 Growth of Japanese flounder larvae at 16 days after hatching fed with rotifers from different growth phases (Redrawn from Tomoda et al. 2006). An *asterisk* indicates significant difference among rotifer growth phases ($n = 3$, Scheffe's F, $p < 0.05$)

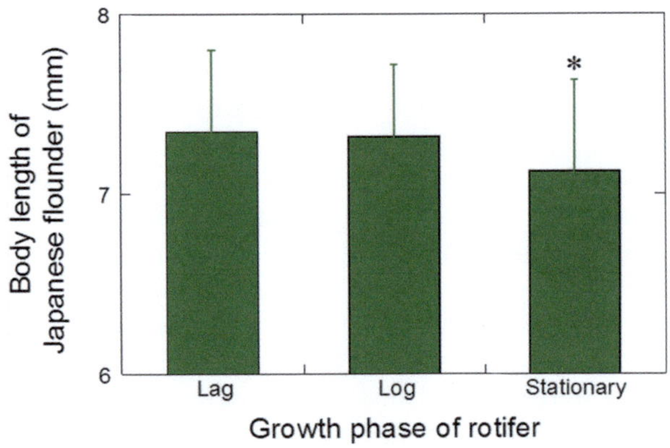

ary phase showed inferior growth than those fed with rotifers in the other growth phases ($n = 3$, Scheffe's F, $p < 0.05$; Fig. 5.2). A similar finding was reported in another marine fish, Japanese flounder *Paralichthys olivaceus*, which is the major target species for both aquaculture and stock enhancement program in Japan (Tomoda et al. 2005). They prepared L-type rotifer at three different growth phases in the same manner of Tomoda et al. (2004): lag phase at day 2 (322 ind./mL), log phase at day 4 (>1000 ind./mL), and stationary phase (4900 ind./mL) at day 8. Japanese flounder larvae were reared from 3 to 16 days after hatching fed with rotifers at three different growth phases. Significantly lower growth and development were detected when they were fed with rotifers in stationary phase ($n = 3$, Scheffe's F, $p < 0.05$; Fig. 5.3).

Furthermore, the growth phase of rotifers when they were fed to marine fish larvae was associated with malformations in fish juveniles. Tomoda et al. (2006)

Table 5.1 Occurrence of malformations in the juvenile Japanese flounder (n = 200) fed with rotifers at different growth phases during the larval stage

	Malpigmentation (%)		
Growth phase of rotifers	Pseudoalbinism on the ocular side	Hypermelanosis on the blind side	Vertebral deformity (%)
Lag	2.5	71.5	41.0
Log	0.5*	23.5*	48.5
Stationary	2.5	66.5	54.5*

Modified from Tomoda et al. (2008)
An asterisk denotes significant difference at $p < 0.05$

reared Japanese flounder larvae by feeding them rotifers at three different growth phases until 24 days after hatching same as their previous protocol (Tomoda et al. 2005) and then fed Japanese flounder larvae with HUFA-enriched *Artemia* and artificial diets until 55 days after hatching when the fish became juvenile and completely settled. Japanese flounder juveniles that were fed with rotifers at the stationary phase during the larval stage had more frequent malformations than those fed with rotifers at the other growth phases (Table 5.1). A noteworthy finding is that the feeding conditions at larval period affected the quality of juveniles in terms of morphological development, even though fish were fed with artificial diet containing sufficient nutritional elements for a longer period than they were fed with rotifers.

These findings by Tomoda et al. (2004, 2005, 2006) clearly demonstrated that the growth phase of rotifers affects the quality of marine fish larvae and that the performance of fish larvae is better when they are fed with rotifers in exponential growth phase. In batch culture, rotifers are generally harvested in the stationary growth phase, causing the decline of their biological activities as well as of their nutritional quality as live food (Koiso and Hino 2002). Therefore, ensuring that the nutritional requirements of fish larvae are met necessitates that the growth phase and condition of rotifers should be carefully examined. The quality of marine fish larvae has been shown to be influenced by the population growth phase of rotifers when they are fed to fish larvae. This finding indicates that the viability of rotifers affects the quality of marine fish larvae. Since the viability of rotifers varies according to their culture conditions, evaluating the viability of rotifers reared using different culture methods and comparing the quality of fish larvae fed with these rotifers are necessary. Unfortunately, only presumptive evidences are available in this regard.

5.3 Distribution and Behavior of Rotifers in Rearing Tanks for Marine Fish Larvae

Fish culturists check routinely the rotifer numbers (density) in the rearing tanks of fish larvae. Many Japanese hatcheries adjust the rotifer density in the larviculture tanks at 5–20 ind./mL, and many of them carefully observe the distribution of fish larvae in the rearing tanks and adjust the rotifer density as targeted level around the water column where larval fish density is high. These culturists claim that observing the rotifer distribution and behavior in the fish rearing tank is compatible to the initiation of active feeding by fish larvae. Rotifers and marine fish larvae are believed to be planktonic and to be easily distributed in the water column by the flow in the rearing tank, although not many quantitative researches have been conducted on the distribution and behaviors of rotifers in the larviculture tanks. Thus, recent findings on the relationship between flow field in the larviculture tanks and distribution of rotifers and fish larvae will be reviewed in this section.

A series of larviculture experiments using seven-band grouper *Epinephelus septemfasciatus* was conducted in order to clarify the optimal flow in the rearing tank and distribution of rotifers and fish larvae. In these studies (Shiotani et al. 2003, 2005; Sakakura et al. 2006), cylindrical tanks (1.0 and 100 kL) were used, and the aspect ratio (AR; water depth/water surface diameter) of these tanks was constant level at about 1.0. One aerator (air stone) was set at the center of the bottom of the tank, and different aeration rates (0, 50, 200, and 1000 mL/min) were proposed. Then, flow field was measured by acoustic Doppler velocity meter in a vertical section on a radius of the rearing tank, distribution of rotifers (*Brachionus rotundiformis*) and fish larvae in the water column was observed, and survival rates of fish larvae 10 days after hatching were compared. It is confirmed that an aerator placed at the bottom center of the rearing tank created the upward flow with air bubble, radiated horizontal flow under the free water surface toward the side wall from the center of tank, downward flow near the side wall, and horizontal flow near the bottom toward the center of rearing tank from the side wall in rearing tank (Shiotani et al. 2003, 2005). This velocity profile was similar to the all aeration rates except for flow speed which increased according to the aeration rate. Aeration at 200 mL/min produced the highest survival among four aeration rates. Measurements of stationary flow showed that the stationary flow in the rearing tank was vertical (Fig. 5.4) and the horizontal circulation was almost negligible. At the aeration rate 200 mL/min, vertical flow velocity above an aerator was about 8 cm/s and the maximum flow velocity at free water surface was 6 cm/s. Fish larvae formed dense patchiness beneath the free water surface; however, rotifers were distributed evenly in the water column when aeration was provided. Those flows around the aerator may be still strong considering the body size of the seven-band grouper larvae (less than 2 mm at hatching). However, no aeration and weak aeration (50 mL/min) tanks showed quite a low survival and lower feeding activity. These results indicate that relatively low flow in the rearing tank reduces the chance to capture prey organism for grouper larvae and/or causes some physical stress such as reducing swimming activity of larvae. Strong aeration may cause high physical stress by both strong current and air bubbles especially around the vertical flow above the aerator.

The proportion of a rearing tank influences both the flow field and performance of marine fish larvae. Ruttanapornvareesakul et al. (2007) introduced two marine fish species larvae, seven-band grouper and devil stinger *Inimicus japonicus*, into commercially available tanks with three different proportions. These tanks were set with the same water volume and aeration rate, but different aspect ratios (AR),

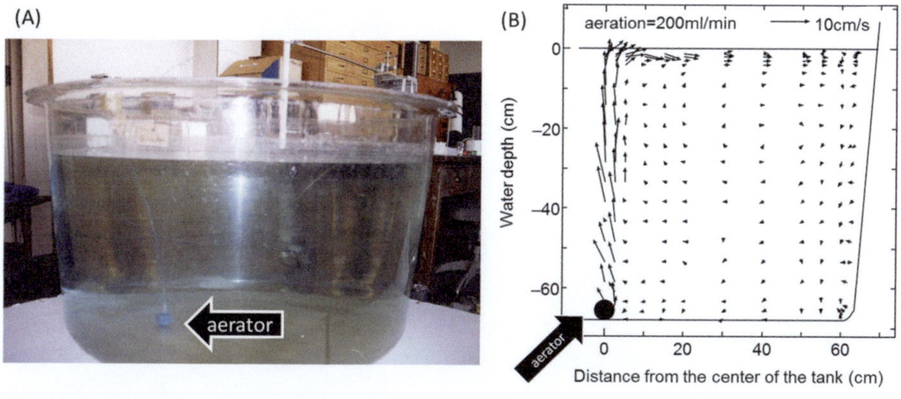

Fig. 5.4 Experimental tank for the measurements of flow field (**a**) and the current velocities in the experimental tank at an aeration rate 200 mL/min (**b**) (Redrawn from Shiotani et al. 2005)

where greater AR indicates that a tank has lower water surface with greater water depth. Survival of larvae of both the species in a tank having an AR greater than 2.0 was found to be significantly higher than that observed in a tank with a lower AR, but growth rates were the same in both the tanks. Fish reared in tanks with a high AR showed less physical stress as determined by enzyme activities (Ruttanapornvareesakul et al. 2010). The reason for the greater survival rate of larvae in a tank with an AR greater than 2.0 was speculated to be that the chances of larvae being captured by the surface tension were reduced because of the greater speed of the larvae at the water surface, and thus the number of deaths related to surface tension were reduced (Ruttanapornvareesakul et al. 2007, 2010). Recently, Sumida et al. (2013) quantified and visualized the flow field in the tanks with different ARs. Flow field in a vertical cross section of a circular tank changed from a single-pair vortex system, which is generally accepted for the flow pattern in a rearing tank with one air stone, to two-pair vortex systems as the value of AR changed from 1.0 to 2.0 (Fig. 5.5). Further, aeration had a weak effect on the changes in the vortex pair system. Larvae in tanks with high ARs might be subjected to either of the two-pair vortex structures as shown in Fig. 5.5; thus, the chance of them to attach to the free surface might be reduced compared to that in a tank with an AR of less than 2.0. The distribution of rotifers might follow a similar pattern, and their physiological status might change according to AR of the rearing tanks. In the future, measurements need to be performed in tanks accommodating both rotifers and larvae to carefully determine how they move and change physiologically in relation to the two-pair vortex structures. As described above, flow field in the cylindrical tanks generated by an aerator has been investigated. However, flow fields in the rectangular tanks, which are used widely and have many different proportions, are not clarified yet. We have started investigating the flow field in the rectangular tanks, which is very complicated comparing to a cylindrical tank and will be reported in the near future.

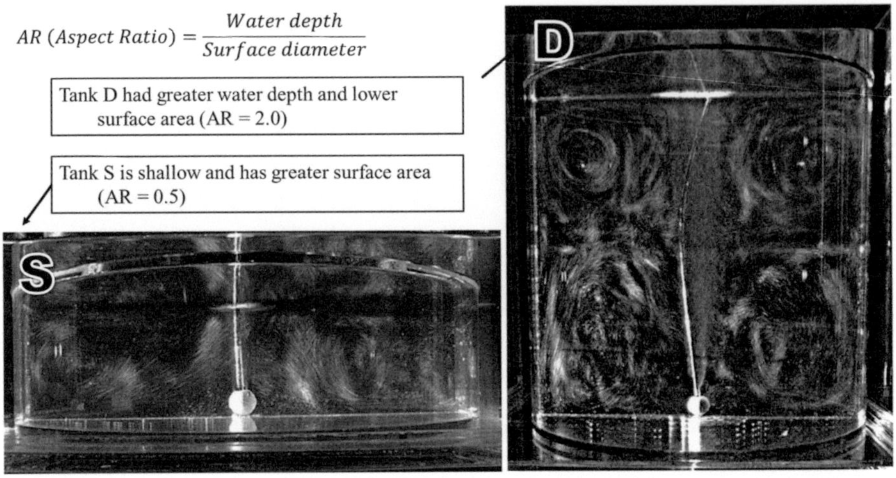

Fig. 5.5 Flow visualization of overall flow pattern in vertical cross section at (S) AR = 0.5 and (D) AR = 2.0, respectively (Modified from Sumida et al. 2013). In S-tank (AR = 0.5), large vortex structures exist inside the tank; one pair symmetrically arranged left and right. Two pairs are seen in D-tank, with one pair in the upper region and another in the lower region

A novel approach patented by Yogo (2008) is considerably successful in the larviculture of several marine fishes. In this rearing system, a salinity gradient is formed in a rearing tank, and the fish are allowed to choose where to stay (Fig. 5.6), unlike that reported by former studies that investigated the effects of salinity on fish by introducing them to different salinity conditions. In the case of devil stinger that becomes demersal after the transition from larva to juvenile, a 3-week rearing trial revealed that its growth and development were synchronized in a tank with salinity gradients with little standard deviation compared to that in an aerated tank (Sakakura et al. 2014). In the tank with salinity gradient, newly hatched devil stinger larvae were distributed in the surface layer (22 ppt), and after mouth opening, they were distributed both on the surface and in the middle layer (22–30 ppt). On day 21, most fish were distributed in the bottom layer (34 ppt) as settled juveniles. However, rotifers were mainly distributed on the surface layer (22–25 ppt), and adjusting rotifer density to the layer where fish swam was necessary. The viability of rotifers was presumed to be higher at the surface layer (low-salinity water) than that at the middle or bottom layers (high salinity water), considering that rotifers are naturally adjusted to low-salinity conditions.

Phototaxis of *Brachionus plicatilis* species complex is affected by light wavelength and intensity (Kim et al. 2014a). The maximum absorbance of wavelength at the pigmented area of the eyespot in rotifers induced strong positive phototaxis as well as reproduction of rotifers (Kim et al. 2014a). Kim et al. (2014b) compared the population growth and phototaxis of rotifers fed with *Nannochloropsis oculata* and baker's yeast (*Saccharomyces cerevisiae*) that lacks vitamin A. Rotifers cultured with baker's yeast showed significantly lower population growth and reduced pig-

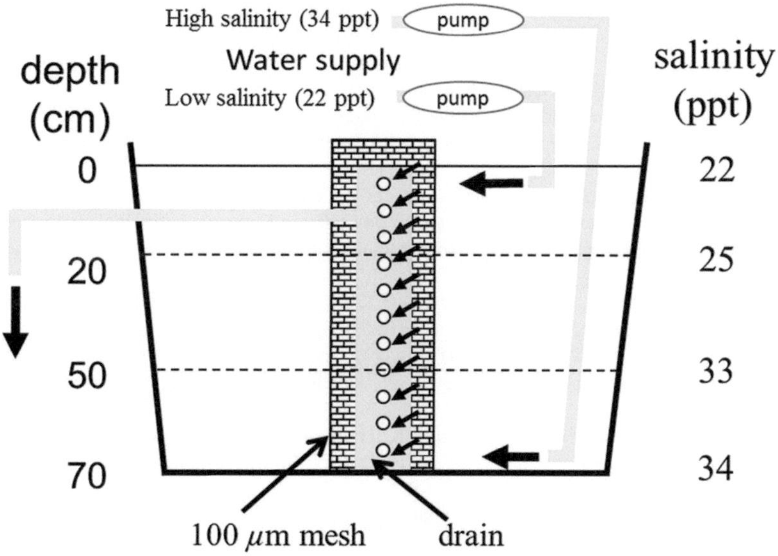

Fig. 5.6 Outline of a salinity gradient tank. (**a**) Schematic diagram of a salinity gradient tank (Patented by Yogo 2008; Redrawn from Sakakura et al. 2014). A salinity gradient was formed by continuously pumping brackish water (22 ppt) from the surface and seawater (34 ppt) from the bottom at the same time at a flow rate of about 350 mL/min for each salinity water into the salinity gradient tank. Water at every depth is discharged from the drain located at the center of the tank. (**b**) A photo showing the salinity gradient tank, which has three inlet with three different salinities (Yamane et al. 2015)

mented area and sensitivity of the rotifer eyespot, resulting in weak phototaxis compared to that in rotifers fed with *N. oculata*. These findings indicate that food source of rotifers affects both the viability and behavior of rotifers, i.e., the behavior of rotifers can be manipulated by food sources. Since marine fish larva is a visual feeder, manipulating light condition is also an important environmental factor for larviculture. Sakakura et al. (2006) found that patchiness by *E. septemfasciatus* larvae under the free water surface became positively correlated with the light intensity of water surface when fish larvae shift into the post-flexion stage. Future studies should focus on manipulating the distribution and behavior of rotifers in a fish rearing tank by modifying water flow, food source, and light conditions, in order to enhance rotifer viability and feeding incidence of marine fish larvae.

References

Bell JG, McEvoy LA, Estevez A, Shields RJ, Sargent JR (2003) Optimizing lipid nutrition in first feeding flatfish larvae. Aquaculture 227:211–220

Conceição LEC, Yúfera M, Makridis P, Morais S, Dinis MT (2010) Live feeds for early stages of fish rearing. Aquac Res 41:613–640

Hagiwara A, Gallardo WG, Assavaaree M, Kotani T, de Araujo AB (2001) Live food production in Japan; recent progress and future aspects. Aquaculture 200:111–127
Hagiwara A, Suga K, Akazawa A, Kotani T, Sakakura Y (2007) Development of rotifer strains with useful traits for rearing fish larvae. Aquaculture 268:44–52
Hamre K, Yúfera M, Rønnestad I, Boglione C, Conceição LEC, Izquierdo M (2013) Fish larval nutrition and feed formulation: knowledge gaps and bottlenecks for advances in larval rearing. Rev Aquac 5:S26–S58
Itoh T (1960) On the culture of mixohaline rotifer *Brachionus plicatilis* O. F. Müller in the sea water. Rep Fac Fish Prefectural Univ Mie 3:708–740. (in Japanese with English abstract)
Izquierdo MS (1996) Essential fatty acid requirements of cultured marine fish larvae. Aquac Nutr 2:183–191
Izquierdo MS, Watanabe T, Takeuchi T, Arakawa T, Kitajima C (1989) Requirement of larval red seabream *Pagrus major* for essential fatty acids. Nippon Suisan Gakkaishi 55:859–867
Kim H-J, Sawada C, Hagiwara A (2014a) Behavior and reproduction of the rotifer *Brachionus plicatilis* species complex under different light wavelengths and intensities. Int Rev Hydrobiol 99:151–156
Kim H-J, Sawada C, Rhee J-S, Le J-S, Suga K, Hagiwara A (2014b) Nutritional effects on the visual system of the rotifer *Brachionus plicatilis* sensu stricto (Rotifera: Monogononta). J Exp Mar Biol Ecol 460:177–183
Koiso M, Hino A (2002) Biological examinations on growth stagnation mechanism in the mass-producing batch culture of the rotifer, *Brachionus plicatilis*. Aquac Sci 50:197–204
Kotani T, Genka T, Fushimi H, Hayashi M, Dierckens K, Sorgeloos P (2009) Effect of cultivation methods on nutritional enrichment of euryhaline rotifer *Brachionus plicatilis*. Fish Sci 75:975–984
Lubzens E (1987) Raising rotifers for use in aquaculture. Hydrobiologia 147:245–255
Mæhre HK, Hamre K, Elvevoll EO (2013) Nutrient evaluation of rotifers and zooplankton: feed for marine fish larvae. Aquac Nutr 19:301–311
Rainuzzo JR, Reitan KI, Olsen Y (1997) The significance of lipid at early stages of marine fish: a review. Aquaculture 155:103–116
Ruttanapornvareesakul Y, Sakakura Y, Hagiwara A (2007) Effect of tank proportions on survival of seven-band grouper *Epinephelus septemfasciatus* (Thunberg) and devil stinger *Inimicus japonicus* (Cuvier) larvae. Aquac Res 38:193–200
Ruttanapornvareesakul Y, Sakakura Y, Hagiwara A (2010) Screening of enzyme activity for assessing the condition of larvae in the seven-band grouper *Epinephelus septemfasciatus* and devil stinger *Inimicus japonicus*. Fish Sci 76:295–304
Sakakura Y, Shiotani S, Chuda H, Hagiwara A (2006) Improvement of the survival in the seven-band grouper, *Epinephelus septemfasciatus*, larvae by optimizing aeration and water inlet in the mass-scale rearing tank. Fish Sci 72(5):939–947
Sakakura Y, Andou Y, Tomioka C, Yogo S, Kadomura K, Miyaki K, Hagiwara A (2014) Effects of aeration rate and salinity gradient on the survival and growth in the early life stages of the devil stinger *Inimicus japonicus*. Aquac Sci 62:99–105
Shiotani S, Akazawa A, Sakakura Y, Chuda H, Arakawa T, Hagiwara A (2003) Measurements of flow field in rearing tank of marine fish larvae: a case study of the seven band grouper *Epinephelus septemfasciatus*. Fish Eng 39:205–212. (in Japanese with English abstract)
Shiotani S, Hagiwara A, Sakakura Y, Chuda H (2005) Estimation of flow in a rearing tank of marine fish larvae by simplified numerical computation – a case of two-dimensional flow. Aquac Eng 32:465–481
Sumida T, Kawahara H, Shiotani S, Sakakura Y, Hagiwara A (2013) Observations of flow patterns in a model of a marine fish larvae rearing tank. Aquac Eng 57:24–31
Takeuchi T (2001) A review of feed development for early life stages of marine finfish in Japan. Aquaculture 200:203–222
Takeuchi T (2014) Progress on larval and juvenile nutrition to improve the quality and health of seawater fish: a review. Fish Sci 80:389–403

Tocher DR (2010) Fatty acid requirements in ontogeny of marine and freshwater fish. Aquac Res 41:717–732

Tomoda T, Koiso M, Kuwada H, Chern J-N, Takeuchi T (2004) Dietary value of marine rotifer *Brachionus plicatilis* in different population growth stages for larval red seabream *Pagrus major*. Nippon Suisan Gakkaishi 70:573–582

Tomoda T, Koiso M, Kuwada H, Chen J-N, Takeuchi T (2005) Dietary value of marine rotifer *Brachionus plicatilis* at different population growth stages for larval Japanese flounder *Paralichthys olivaceus*. Nippon Suisan Gakkaishi 71:555–562

Tomoda T, Koiso M, Kuwada H, Chen J-N, Takeuchi T (2006) Larval development and occurrence of morphological abnormalities in hatchery-reared Japanese flounder *Paralichthys olivaceus* larvae fed marine rotifer *Brachionus plicatilis* at different population growth stages. Nippon Suisan Gakkaishi 72:725–733

Tomoda T, Koiso M, Shima Y (2008) Temperature effects of rotifer *Brachionus plicatilis* mass culture on rearing efficiency of larval Japanese flounder *Paralichthys olivaceus*. Nippon Suisan Gakkaishi 74:625–635

Tsukamoto K, Kuwada H, Uchida K, Masuda R, Sakakura Y (1999) Fish quality and stocking effectiveness: behavioral approach. In: Howell BR, Moksness E, Svåsand T (eds) Proceedings of first international symposium on stock enhancement and sea ranching. Blackwell Science, London, pp 205–218

Watanabe T (1993) Importance of docosahexaenoic acid in marine larval fish. J World Aquacult Soc 24:152–161

Watanabe T, Izquierdo MS, Takeuchi T, Satoh S, Kitajima C (1989) Comparison between eicosapentaenoic and docosahexaenoic acids in terms of essential fatty acid efficacy in larval red seabream. Nippon Suisan Gakkaishi 55:1625–1640

Yamane H, Yoshida A, Yamazaki H, Sakiyama K, Kawabata Y, Sakakura Y (2015) Changes in behavior and feeding habits of hatchery-reared tiger puffer *Takifugu rubripes* juveniles. Aquac Sci 63:141–149. (in Japanese with English abstract)

Yogo S (2008) Management methods of artificial seedlings. Japanese Patent 4091965

Chapter 6
Use of Freshwater *Brachionus* for Aquaculture

Yuka Ogata

Abstract Due to the rising demand for aquaculture and fish seed production, the utilization of freshwater *Brachionus* rotifer was discussed. One tropical strain, *Brachionus angularis*, isolated from Southeast Asia was studied, including its biological characteristics, as well as results of feeding tests to indigenous freshwater fish in comparison to other larval foods. Also, ways to isolate rotifer strains from a natural water body were described.

6.1 Introduction

Freshwater rotifers have the long history of being investigated as first food in freshwater seed production (e.g., Groeneweg and Schliiter 1981; Awaïss et al. 1992), yet relatively less reported compare to brackish and saltwater rotifers. Pure culture (single-species culture) of freshwater *Brachionus* sp. is usually unstable and difficult to maintain compared to marine *Brachionus* sp., mainly due to the absence of ionic compound in freshwater which serves as buffer in seawater environment. Also, because of this, freshwater rotifer cultures are easily contaminated with various bacteria and protozoan which suppress rotifer population growth especially in large-scale culture. The cosmopolitan species such as *Brachionus calyciflorus*, *B. rubens*, and *B. angularis* are commonly found in eutrophic waters in tropical to temperate climate areas. These rotifer species have considerably similar biological characteristics to *B. plicatilis* sp. complex; thus, utilizing freshwater *Brachionus* sp. for freshwater seed production is recommended. This chapter will introduce one strain of *B. angularis* which shows stable population growth in artificial culture condition. Also, described are its biological characteristics, as well as the result of feeding tests to fish larvae in comparison to other live foods. In addition, ways to isolate these strains from a natural freshwater body are discussed from the perspective of using this species in culturing indigenous fishes.

Y. Ogata (✉)
Department of Aquaculture and Marine Environment, Niigata Prefectural Fisheries and Marine Research Institute, 13098-8 Sannocho, Igarashi, Nishi-ku, Niigata 950-2171, Japan
e-mail: ogata.yuka@pref.niigata.lg.jp

A. Hagiwara, T. Yoshinaga (eds.), *Rotifers*, Fisheries Science Series,
DOI 10.1007/978-981-10-5635-2_6

6.2 Biological Characteristics of *B. angularis*

B. angularis is a small-sized (lorica length of adult amictic female, <150 μm), cosmopolitan (Yin and Niu 2008; Leutbecher 2000; Gilbert and Burns 1999; Pourriot and Rougier 1997), freshwater rotifer species. Its combination of asexual and sexual reproduction, round shape, and microalgal diet is similar to those of the brackish water rotifer *Brachionus plicatilis* sp. complex, which is an important species in the field of saltwater aquaculture as a starter food for larval rearing. *B. angularis* is smaller than *B. plicatilis*, and only few studies have investigated the use of *B. angularis* as a live feed for larval fish so far.

B. angularis Laos strain was originally isolated for aquaculture use by the researchers from the University of Tokyo. The strain has a pot-shaped lorica similar to that of other *Brachionus* species (Fig. 6.1). Immediately after hatching, females have an oblong lorica; subsequently, both the edges of the lorica gradually inflate during growth. The lorica size in mature females is considerably smaller than that in other strains of *B. angularis* (Table. 6.1). Mature females generally carry one to three parthenogenetic egg(s) and occasionally carry approximately five male eggs. Parthenogenetic eggs are elliptical and larger than male eggs, which are spherical. Resting eggs are oval and have cavities similar to those in the eggs of *B. plicatilis*; resting eggs are larger than parthenogenetic eggs.

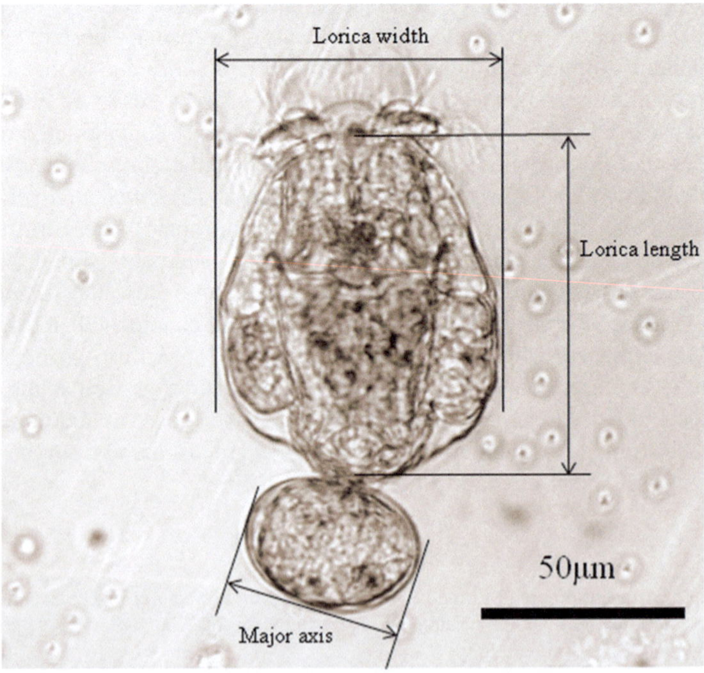

Fig. 6.1 *Brachionus angularis* Laos strain (Source: Ogata et al. 2011)

Table 6.1 Size comparison of *B. angularis* Laos strain to the other species and strains from different origins

Species/Origin	Lorica length (n)	Lorica width (n)	Major axis of parthenogenetic egg	Reference
B. angularis/Laos	86.0 ± 4.9 (20)	75.6 ± 5.7 (20)	60.3 ± 3.1 (83.3 ± 3.3)	Present study
B. angularis/China	130 ± 7	115 ± 7	–	Yin and Niu (2008)
B. angularis/Germany	120–140	–	–	Leutbecher (2000)
B. angularis/New Zealand	122	–	–	Gilbert and Burns (1999)
B. angularis/France	127.8 ± 5.9	–	–	Pourriot and Rougier (1997)
Brachionus[a] (SS-type) Thai strain	179 ± 4	–	–	Hagiwara et al. (1995)
B. plicatilis[a] (S-type)	212 ± 20	171.9 ± 13.0	–	Fu et al. (1991)
B. plicatilis (L-type)	275 ± 24	214.3 ± 17.3	–	Fu et al. (1991)

Source: Ogata et al. (2011)
[a]Subsequently classified as *B. plicatilis* sp. complex

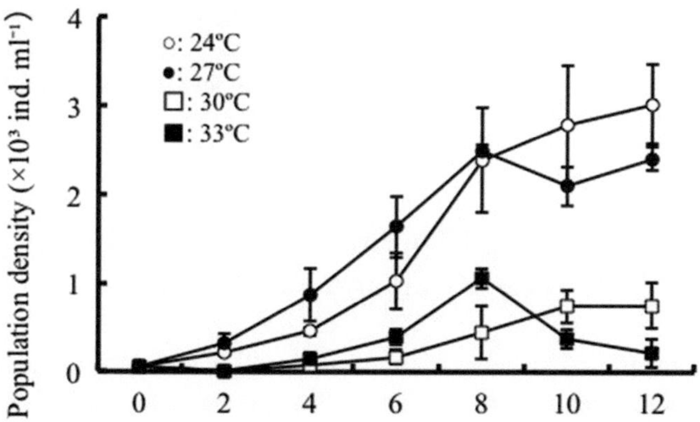

Fig. 6.2 Population growth of *B. angularis* Laos strain at various temperatures after feeding 10 × 10^6 *Chlorella* cells·ml^{-1} (Source: Ogata et al. 2011)

The optimum culture condition for this strain was a temperature of approximately 24–27 °C (Fig. 6.2); under these conditions, the rotifer density reaches >2000 ind/ml within 10 days in small-scale (100 ml) batch cultures. However, *B. angularis* Laos strain did not grow well at higher temperatures (30 °C and 33 °C), even though the strain was isolated from a tropical monsoon region (Vientiane, Laos) where the maximum air temperature can exceed 30 °C for 10 months in a year

(Sipaseuth et al. 2009). The organisms are generally adapted to native climatic conditions; however, this discrepancy might be because of the deterioration of water quality owing to the accumulation of uneaten prey and excretory products of rotifers; this deterioration was accelerated at higher temperatures in a closed environment.

Brachionus strains should be cultured under various conditions after isolation. Galkovskaja (1987) emphasized the importance of determining the temperature tolerance ranges of candidate species in order to optimize culture conditions for maximum production. In addition, at a fixed temperature, the population growth rate of *Brachionus* varies according to the culture volume. The results of test culture in different vessel sizes can allow the determination of the growth characteristics of the isolated strain. Cultures of a single or very few individuals of *Brachionus* sp. are often used for analytical study on life history (Athibai and Sanoamuang 2008; Fussmann et al. 2007; Somamihardja and Bart 2008; Yoshinaga et al. 2003). These culture procedures are also useful for investigating the optimum conditions of *Brachionus* for aquaculture.

The use of small rotifers for use in larviculture was reported by Hagiwara et al. (1995) and Wullur et al. (2009); these studies were performed on brackish water rotifers, SS-type *Brachionus* and *Proales similis*, respectively. The small size of *B. angularis* should be very advantageous for feeding small-mouthed freshwater fish larvae. In marine larviculture, the use of small rotifers as food significantly improves the initial feeding performance of fish larvae and increases their survival rates (Polo et al. 1992; Planas and Cunha 1999). Planas and Cunha (1999) suggested that rearing fish larvae up to the stage of exogenous feeding from the mouth opening can be improved by supplying small rotifers as food. Feeding *B. angularis* to fish larvae might similarly improve their survival and performance in freshwater larviculture.

6.3 Results of Feeding Experiment Performed Using *B. angularis* Laos Strain and Fish Larvae

The result of feeding experiments conducted on silver barb, *Hypsibarbus malcolmi*, the Laotian indigenous cyprinid, by using *B. angularis* Laos strain as a live food is shown in Fig. 6.3. In the experiment, two 20 l polycarbonate tanks were prepared, and 300 larvae (3 days after hatching (DAH) in preflexion larval period; Ogata et al. 2010) were introduced in each tank. The larvae in one tank were reared until 12 DAH and were fed with *B. angularis* Laos strain at 5–10 ind/ml. Population density of the rotifers was maintained twice daily (at 0900 and 1500) to maintain the density range by adding a decreased number of rotifers according to consumption by the larvae. The larvae in another tank were maintained without food through the end of the experiment. All larvae were sampled at 12 DAH; the number of surviving individuals was counted, and total length (TL) was measured. Each treatment was

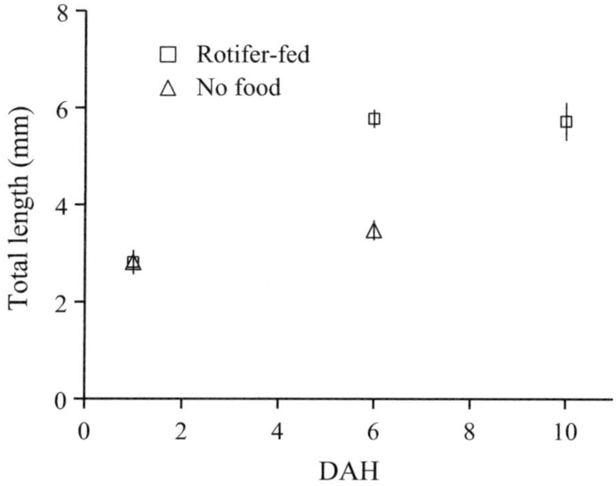

Fig. 6.3 Growth of *H. malcolmi* larvae with and without *B. angularis* Laos strain. *Bars* represent standard deviation. Larvae with no food supply (△) died by 10 DAH

Table 6.2 Growth and survival of *H. malcolmi* larvae with and without *B. angularis*

Days after hatch		No food	Rotifer-fed
3	Total length (SD) (mm)	2.8 (0.3)	2.8 (0.3)
6	Total length (SD) (mm)	3.5 (0.2)[a]	5.8 (0.2)[b]
	Survival rate (%)	100	100
10	Total length (SD) (mm)	No data	5.8 (0.4)
	Survival rate (%)	0	100

Values with different superscript letter are significantly different ($p < 0.05$)

applied in triplicate. Larvae fed with *B. angularis* showed 100% survival at 10 DAH (Table 6.2). These results confirm that *H. malcolmi* larvae feed on *B. angularis* Laos strain; the larvae increased in body size after feeding, indicating that *B. angularis* not only has a morphologically edible size as a live food but also is nutritionally valuable.

The body growth of *H. malcolmi* larvae fed with *B. angularis* in the beginning and in sequence followed by other artificially prepared common live foods, and as comparison fed with wild zooplanktons in other treatment is shown in Fig. 6.4 and Table 6.3. Rearing tanks were filled with 30 l of freshwater, and at 3 DAH, the larvae were introduced into each tank at 10 ind/l. The larvae were reared under ambient water temperatures ranging from 25.4 to 30.0 °C until 28 DAH under two different feeding regimes. Larvae in the first tank were fed with *B. angularis* Laos strain (5 ind/ml, 2–12 DAH), *Artemia* sp. nauplii (body length approximately 400 μm, 0.5 ind/ml, 6–19 DAH), copepods (size range 100–300 μm, 0.1 ind/ml, 8–19 DAH), *Moina* spp. (size range 100–500 μm, 0.1 ind/ml, 20–28 DAH), and catfish pellets (particle diameter 300–500 μm, 0.1 g/day/tank, 20–28 DAH). Larvae in the second tank were fed with planktons (size range 10–500 μm) obtained from aquaculture ponds in Vientiane, Laos. The population density of *B. angularis* and other artificially prepared live food organisms in the first tank was maintained twice daily

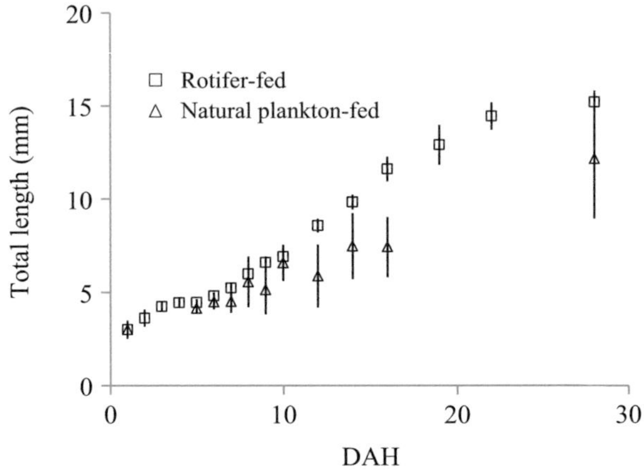

Fig. 6.4 Growth of *H. malcolmi* larvae fed with *B. angularis* Laos strain and natural planktons. *Bars* represent standard deviation. No sample was taken in natural plankton-fed tanks on 19 and 22 DAH because few survivors are in the tank

Table 6.3 Survival and growth of *H. malcolmi* larvae fed with *B. angularis* and natural plankton on 28 DAH

	Rotifer-fed	Natural plankton-fed
Survival (%)	94^{a}	6^{b}
Total length (SD) (mm)	$15.2\ (0.6)^{a}$	$12.2\ (3.2)^{b}$

Values with different superscript letter are significantly different at $p < 0.05$

(at 0900 and 1500) to maintain the density range by adding a decreased number of the organisms according to consumption by the larvae. Natural planktons (mixture of rotifers, copepods, and cladocerans) were filtered and sorted by size (small, 10–100 μm; large, 100–500 μm) for the second tank. Larvae in that tank were fed once daily with 14 ind/ml of small planktons and 0.6 ind/ml of large planktons. At 1–10, 12, 14, 16, 19, 22, and 28 DAH, the larvae were sampled from the first tank, and the number of surviving individuals was counted. TL and body depth (BD) were measured to the nearest 0.1 mm. Final survival rate was calculated from the number of surviving individuals in the second tank at the end of the experiment. Each treatment was applied in duplicate. The survival and TL of larvae fed with *B. angularis* were high (99%, 12.9 mm at 19 DAH and 94%, 15.2 mm at 28 DAH, respectively). In contrast, the larvae fed with natural zooplanktons showed low survival (6% at 28 DAH, Table 6.3) and large variation in TL at the end of the experiment (Fig. 6.5).

Ogata and Kurokura (2012) reared *B. splendens* larvae by using *B. angularis* Laos strain, *Paramecia* sp., and *Artemia* as live food sources. In the experiment, rearing tanks were filled with 1500 l of freshwater (maintained 27 ± 1 °C), and at 3 DAH, 22 larvae were introduced into the tank (15 ind/l). The larvae were reared until 18 DAH under four different feeding regimes: (1) no food, (2) *Paramecium* sp. (10 ind/ml until 7 DAH, 20 ind/ml from 8 DAH), (3) *B. angularis* Laos strain (10 ind/ml until 7 DAH, 20 ind/ml from 8 DAH), and (4) *B. angularis* (10 ind/ml until 7 DAH) and *Artemia* sp. nauplii (1 ind/ml from 8 DAH). The density of food animals provided was set at the satiation level of the larvae. Food organism density

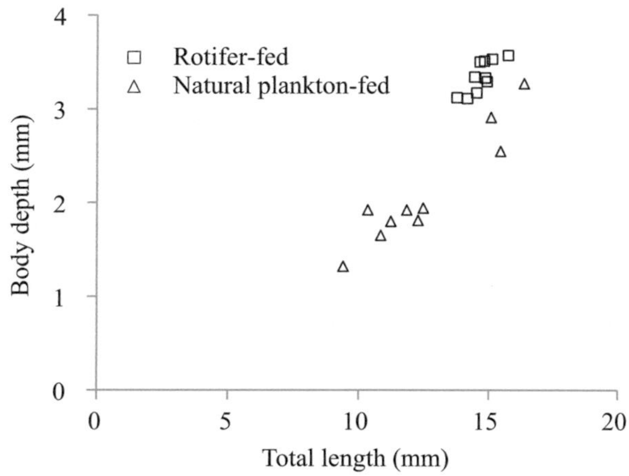

Fig. 6.5 Total length and body depth of *H. malcolmi* larvae fed with *B. angularis* and natural planktons on 30 days after hatching

was determined twice daily and was adjusted for the decreased number of organisms according to food consumption by the larvae. For more detailed experiment conditions, refer Ogata and Kurokura (2012).

Larvae fed with live food were found to have a significantly higher survival rate (97.5–100%) at 18 DAH than that of the control (unfed) larvae, which died by 12 DAH. *B. angularis*-fed larvae were found to grow faster than *Paramecia*-fed larvae. The fastest growth rate was observed in larvae fed with a combination of *B. angularis* and *Artemia*; growth of these larvae increased by 282% by 18 DAH (TL 11.3 ± 1.2 mm) relative to body measurements obtained at 3 DAH. The next fastest growth rate was observed in *B. angularis*-fed larvae, with a 158% increase in growth observed after 18 DAH (TL 7.6 ± 0.5 mm). The *Paramecia*-fed larvae were found to grow by only 54.3% (TL 4.6 ± 0.1 mm) during the same period.

In feeding experiment on Siamese fighting fish, *Betta splendens*, *B. angularis*-fed larvae are superior in terms of growth and survival compared to those fed with other live food. *B. splendens* is a popular ornamental fish native to the Mekong Basin. The larvae have a small body size, and at present, larviculture of this species uses protozoans as food source.

6.4 Recommendation for Culturing Indigenous Species by Using Freshwater *Brachionus*

The use of small rotifers for fish larvae with a small mouth size might widen the variety of culturable freshwater fish, and remarkable potential exists in freshwater aquaculture with new species. Aquaculture production in the Asia–Pacific region is continuously increasing quantitatively and economically and accounts for a considerable share of the total world production (FAO 2007). As of 2011, the largest producer of aquaculture is China, and six Southeast Asian countries (Indonesia, Thailand, Vietnam, Bangladesh, the Philippines, and Myanmar) have been ranked among the top ten aquaculture producers in the world. De Silva et al. (2006) reported that the contribution of the Asian region to global cultured finfish production is 87%. Undoubtedly, recent technological improvements in fish farming have accelerated the increase in production. FAO and NACA (2011) also noted that the Asia–Pacific region is characterized by the predominance of small-scale fish farmers, which was not recognized until recently. The fish farming industry is essential not only for securing a protein source to sustain the growing population in the region but also for providing a source of income and employment. Further, the aquaculture industry contributes to poverty alleviation, especially in developing countries (De Silva et al. 2006).

For example, in Southeast Asia, although various fish species are farmed, most of the total food fish production is dominated by only a few species, mostly cyprinids, tilapias, and catfish, which are freshwater species. In particular, in the countries surrounding the Mekong Basin, which is a rich freshwater resource, many inland fishes are consumed (MRC 2007). There is concern that the natural fishery resources are decreasing (FAO 2007), and the future demand for cultured fishes might increase. African native tilapias and catfish, as well as Eastern Asian native cyprinids, are known to be spreading beyond their native ranges (Athibai and Sanoamuang 2008), and the continuous diffusion of exogenous species might become a controversial issue in the future.

This controversy also includes exotic species, also known as exogenous or alien species, which are generally considered to be not native to the present habitat and ecosystem. Recently, increasing numbers of studies are focusing on the impact of the spread of exotic fish species used for aquaculture in Southeast Asia. Although these studies often credit the function of exotic fish species as a secure food resource, the negative impacts are becoming more apparent. Welcome and Vidthayanon (2003) suggested the leading problems associated with exotic species: (1) environmental disturbance such as digging by common carp and burrowing by crayfish; (2) predation such as the disappearance of many cichlid species by the Nile perch in Lake Victoria, East Africa; (3) competition, such as a dense population of small *Oreochromis niloticus* individuals competing with valuable fishery resources; (4) introduction of diseases such as the koi herpesvirus (Bondad-Reantaso 2004); (5) genetic contamination/hybridization; and (6) co-introduction of nuisance species such as the unintentionally introduced *Pseudorasbora parva* that contaminates imported Chinese carp fry. De Silva et al. (2006) concluded that the most urgent need in Asia is more scientific research on the impact of alien species and the need to develop aquaculture of indigenous species. Freshwater aquaculture in Asian regions is still in its early stage of development. Therefore, there is considerable scope for improvement in these regions. The use of local species for aquaculture as well as investigation of the ecology of these species should be encouraged to allow the culturing of a variety of fish in the future. This might eliminate invasive species and reduce fishing pressure on the wild stock in the area. Technological development in aquaculture should be considered equally important as resource management of wild stocks.

6.5 Isolation of Freshwater *Brachionus*

Two methods are available to isolate a strain of freshwater *Brachionus* and propagate them: one is the intensive method, and the other is the extensive method. The intensive method is recommended when laboratory equipments with temperature control, such as an automated incubator, are available. Further, pure-cultured microalgae such as *Chlorella* sp. should be prepared for isolation and propagation. Extensive method is useful for conditions when an incubator or sterilized working environment is not available and obtaining pure-cultured microalgae is difficult.

For the intensive method, various water samples need to be collected from natural water bodies such as rivers, lakes, and even paddles or artificial ones such as aquaculture ponds. Freshwater *Brachionus* can be often found in artificially fertilized green water for extensive aquaculture. The water samples should then be filtered through fine mesh plankton net (300-μm mesh or lower) to remove larger zooplanktons. Rotifers need to be picked up individually from the filtered samples by using a Pasteur pipette or similar glass pipettes with a very narrow tip, and the rotifers need to be cultured individually in multi-well polystyrene plates (e.g., 96-well plates by Iwaki, Japan) at the desired temperature. Preferably, more number of rotifer individuals need to be picked up. However, in the natural environment, existence and population of zooplanktons fluctuate depending on the conditions such as season and temperature. Therefore, strains can be isolated in dozens over a couple of times or in hundreds at one time depending on the availability. Autoclaved natural freshwater or drinking water filtered using a glass fiber filter (e.g., GF/C; normal rating, 1.2 μm; thickness, 280 μm; Whatman Ltd.) should be used as the culture medium. Freshwater microalgae such as *Chlorella* sp. (e.g., Super Chlorella V-12®; diameter, 3–8 μm; Chlorella Industry Co.) are diluted at a density of around 1.0×10^6 cells/ml and used as food for the rotifers. After a few days of culture, a few dozens of strains with high population growth rates should be selected and cultured for 1 week. Finally, the strain with the highest population growth across all strains should be selected for culturing. The culture volume can be gradually increased for mass culture. For species identification, refer texts such as Wallace et al. (2006) and Suzuki (1999).

In the extensive method, a water tank is filled with natural freshwater and set outdoor. Chicken manure is prepared in a rice bag and floated in the tank. For 150–200 l of water, 1 kg of chicken manure is required to fertilize the water. If natural water such as river water or pond water is used, chances of having large carnival planktons such as copepods in the water are high; hence, organophosphorus insecticides such as Diazinon 60 (5 ppm; 100 ml of the solution contains 60 ml Diazinon) need to be added to remove them. The pesticide is commonly used to control fleas, etc. Direct sunshine might accelerate the *Brachionus* population growth, although it is not necessary. Avoiding rainwater to fall in the tank and thin down the fertilized water requires that the tank should be set under a roof. After several days to a few weeks, microalgae begin to bloom. Subsequently, the population of *Brachionus* increases and shows peak population growth. This procedure should be repeated in order to determine the temperature and days required to have sufficient amount of *Brachionus* to feed the fish larvae. Backward calculation might allow preparing rotifers for fish breeding and larval rearing.

References

Athibai S, Sanoamuang L (2008) Effect of temperature on fecundity, life span and morphology of long- and short-spined clones of *Brachionus caudatus* f. apsteini (rotifera). Int Rev Hydrobiol 93:690–699

Awaiss A, Kestemont P, Micha CJ (1992) An investigation into the mass production of the freshwater rotifer *Brachionus calyciflorus* Pallas. 1. An eco-physiological approach to nutrition. Aquaculture 105:325–336

Bondad-Reantaso M (2004) Trans-boundary aquatic animal diseases: focus on Koi herpes virus (KHV). Aquacult Asia 9:24–28

De Silva S, Nguyen T, Abery N, Amarasinghe U (2006) An evaluation of the role and impacts of alien finfish in Asian inland aquaculture. Aquac Res 37:1–17

FAO (2007) Cultured aquatic species information programme. *Chanos chanos*. Cultured aquatic species information programme. Text by Nelson AL, Marygrace CQ In: FAO Fisheries and Aquaculture Department [online]. Rome. Updated 23 November 2007. [Cited 28 December 2011]. http://www.fao.org/fishery/culturedspecies/Chanos_chanos/en

FAO, NACA (2011) Regional review on status and trends in aquaculture development in Asia-Pacific–2010. Food and Agriculture Organization of the United Nations, Rome

Fu Y, Hirayama K, Natsukari Y (1991) Morphological differences between two types of the rotifer Brachionus plicatilis O. F. müller. J Exp Mar Biol Ecol 151:29–41

Fussmann GF, Kramer G, Labib M (2007) Incomplete induction of mixis in *Brachionus calyciflorus*: patterns of reproduction at the individual level. Hydrobiologia 593:111–119

Galkovskaja GA (1987) Planktonic rotifers and temperature. Hydrobiologia 147:307–317

Gilbert JJ, Burns CW (1999) Some observations on the diet of the backswimmer, *Anisops wakefieldi* (hemiptera: Notonectidae). Hydrobiologia 412:111–118

Groeneweg J, Schliiter M (1981) Mass production of freshwater rotifers on liquid wastes. II. Mass production of *Brachionus rubens* Ehrenberg 1838 in the effluent of high-rate algal ponds used for the treatment of piggery waste. Aquaculture 25:25–33

Hagiwara A, Kotani T, Snell TW, AssavaAree M, Hirayama K (1995) Morphology, reproduction, genetics, and mating behavior of small, tropical marine *Brachionus* strains (rotifera). J Exp Mar Biol Ecol 194:25–37

Leutbecher C (2000) A routine method of DNA-extraction from extremely small metazoans, e.g. single rotifer specimens for RAPD-PCR analyses. Hydrobiologia 437:133–137

MRC (2007) Consumption and the yield of fish and other aquatic animals from the Lower Mekong Basin, MRC technical paper, Lao PDR

Ogata Y, Kurokura H (2012) Use of the freshwater rotifer *Brachionus angularis* as the first food for larvae of the Siamese fighting fish *Betta splendens*. Fish Sci 78:109–1125

Ogata Y, Morioka S, Sano K, Vongvichith B, Eda H, Kurokura H, Khonglaliane T (2010) Growth and morphological development of laboratory-reared larvae and juveniles of the Laotian indigenous cyprinid *Hypsibarbus malcolmi*. Ichthyol Res 57:389–397

Ogata Y, Tokue Y, Yoshikawa T, Hagiwara A, Kurokura H (2011) A Laotian strain of the rotifer *Brachionus angularis* holds promise as a food source for small-mouthed larvae of freshwater fish in aquaculture. Aquaculture 312:72–76

Planas M, Cunha I (1999) Larviculture of marine fish: problems and perspectives. Aquaculture 177:171–190

Polo A, Yufera M, Pascual E (1992) Feeding and growth of gilthead seabream (*Sparus-aurata* L) larvae in relation to the size of the rotifer strain used as food. Aquaculture 103:45–54

Pourriot R, Rougier C (1997) Reproduction rates in relation to food concentration and temperature in three species of the genus *Brachionus* (rotifera). Ann Limnol Int J Limnol 33:23–31

Sipaseuth Basnayake J, Fukai S, Inthapanya P, Changphengxay M (2009) Consistency of genotypic performance of lowland rice in wet and dry season in Lao PDR. Field Crop Res 111:47–54

Somamihardja A, Bart A (2008) Characterization of small-size egg-bearing Thai ss-strain rotifers *Brachionus rotundiformis* and their first offspring. J World Aquacult Soc 39:528–534

Suzuki M (1999) An approach to the identification of the common rotifers. Sanseido, Tokyo

Wallace RL, Snell TW, Ricci C, Nogrady T (2006) Rotifera vol. 1: biology, ecology and systematics. In: Segers H, Dumont HJ (eds) Guides to the identification of the microinvertebrates of the continental waters of the world, 23, 2nd edn. Kenobi Productions/Backhuys Academic Publishing BV, Gent/The Hague

Welcome R, Vidthayanon C (2003) The impact of introductions and stocking of exotic species in the Mekong basin and policies for their control, MRC technical paper No. 9, MRC, Phnom Penh

Wullur S, Sakakura Y, Hagiwara A (2009) The minute monogonont rotifer *Proales similis* de beauchamp: culture and feeding to small mouth marine fish larvae. Aquaculture 293:62–67

Yin XW, Niu CJ (2008) Effect of pH on survival, reproduction, egg viability and growth rate of five closely related rotifer species. Aquat Ecol 42:607–616

Yoshinaga T, Hagiwara A, Tsukamoto K (2003) Life history response and age-specific tolerance to starvation in *Brachionus plicatilis* O. F. Muller (rotifera). J Exp Mar Biol Ecol 287:261–271

Part III
Model Organism

Chapter 7
Life History Variation in Monogonont Rotifers

Claus-Peter Stelzer

Abstract Over the last decades, monogonont rotifers have proven to be excellent experimental models for life history studies. Their short life cycle and the ease of culturing rotifers facilitate experimental studies over one or several generations. Clonal reproduction allows the same genotype to be replicated over an essentially unlimited number of factors and factor combinations. Altogether this has led to a myriad of studies on how different factors can affect life history variables in rotifers. The purpose of this review is to give an overview on the main life history traits of the rotifer life cycle and to summarize the basic sources of variation of these life history traits, i.e., phenotypic plasticity (external factors affecting life history variables of individual genotypes within one generation), genetic variation (genotypic differences among individuals), and other sources of variation (maternal effects, transgenerational effects, developmental stochasticity).

7.1 Introduction: The Life History Traits of Monogonont Rotifers

Life history theory in general deals with the so-called life history traits, which include variables such as size at birth; age and size at maturity; age-specific survival, fecundity, and somatic growth; number, size, and sex of the offspring; and longevity (Roff 1992; Stearns 1992). Altogether these variables determine the fitness of an individual, i.e., its contribution to future generations. Fitness can be quantified as the per-female total number of offspring, R_0, or the intrinsic rate of population increase, *r*, given the age-specific schedules of survivorship and fecundity (Roff 1992; Stearns 1992). According to classical life history theory, fitness is maximized via optimization of the all life history traits in a given environment (biotic and abiotic factors) and under a given set of internal trade-offs and other constraints.

Life history trade-offs can be defined either at the individual level (physiological trade-offs) or at the population level (genetic trade-off). Physiological trade-offs are

---

C.-P. Stelzer (✉)
Research Institute for Limnology, University of Innsbruck, 5310 Mondsee, Austria
e-mail: claus-peter.stelzer@uibk.ac.at

A. Hagiwara, T. Yoshinaga (eds.), *Rotifers*, Fisheries Science Series,
DOI 10.1007/978-981-10-5635-2_7

caused by competitive allocation of limited resources to one life history trait vs. another within a single individual. In clonal organisms, this may become apparent if there is stochastic or environmentally caused variation in allocation to the two traits in different individuals of the same clone. For example, several studies have shown that individuals of a rotifer clone that have elevated reproductive rates in their adult life tend to live shorter (Kauler and Enesco 2011; Sarma et al. 2002), in particular, if they are resource limited (Stelzer 2001). In contrast, genetic trade-offs are caused by (genetically determined) differences in the allocation pattern among individuals, i.e., genotypes of a population. For example, Snell and King (1977) have sampled a large number of clones from a natural *Asplanchna brightwelli* population and demonstrated that clones with a high reproduction early in life had a shorter lifespan than clones with more moderate rates of reproduction.

General concepts of life history theory can be applied to monogonont rotifers; however this is complicated by the fact that the full life cycle encompasses more than one generation and several stages with very distinct development (see Fig. 7.1). Thus, the life history of monogonont rotifer has more variables than usually considered. Figure 7.1 shows a simplified scheme of the monogonont life cycle and lists most relevant life history traits. Since there are both sexual and asexual stages, and because sexual stages can be different (unfertilized females, fertilized females, males), each stage can be characterized, e.g., by its stage-specific survival and fecundity schemes (Chen et al. 2005; Xi et al. 2001). Additionally the intersections between these stages give rise to new life history traits, such as mixis propensity (transition from asexual to sexual cycle) or diapause (transition from sexual to asexual life cycle). Another interesting variable in the rotifer life cycle is the "fertilization threshold," i.e., the maximum age to which a mictic female can be fertilized by a male, which determines if a female will become a male producer or a female (diapausing egg) producer (Gomez and Serra 1996). Thus the fertilization threshold plays a pivotal role in sex allocation in monogonont rotifers (Aparici et al. 1998).

In the context of life history theory, fitness in monogonont rotifers includes all components of the sexual and asexual life cycle. This means that fitness is not determined solely by asexual population growth (r), but also by the timing and degree of switching to the sexual cycle (i.e., the threshold for mixis induction and the maximum mixis ratio), because these factors ultimately determine the contribution to a resting egg bank by a specific genotype. Models on the optimization of the full rotifer life cycle thus typically include all these components (e.g., Serra and King 1999), and some studies quantified fitness by the number of diapausing eggs produced by a clone under given environmental conditions (e.g., Campillo et al. 2009).

The trait "mixis propensity," which reflects the tendency of a clone to invest into sexual reproduction, deserves further notion because it is *the* central life history trait in the monogonont life cycle, and it distinguishes rotifers from most other animals. Mixis can be induced by several factors, including photoperiod, alpha-tocopherol, or crowding chemicals released by the rotifers at high population densities (Nogrady et al. 1993; Schröder 2005). In *B. plicatilis* the mixis factor has been characterized as a 39 kD water-soluble protein (Snell et al. 2006). In this review I will focus on

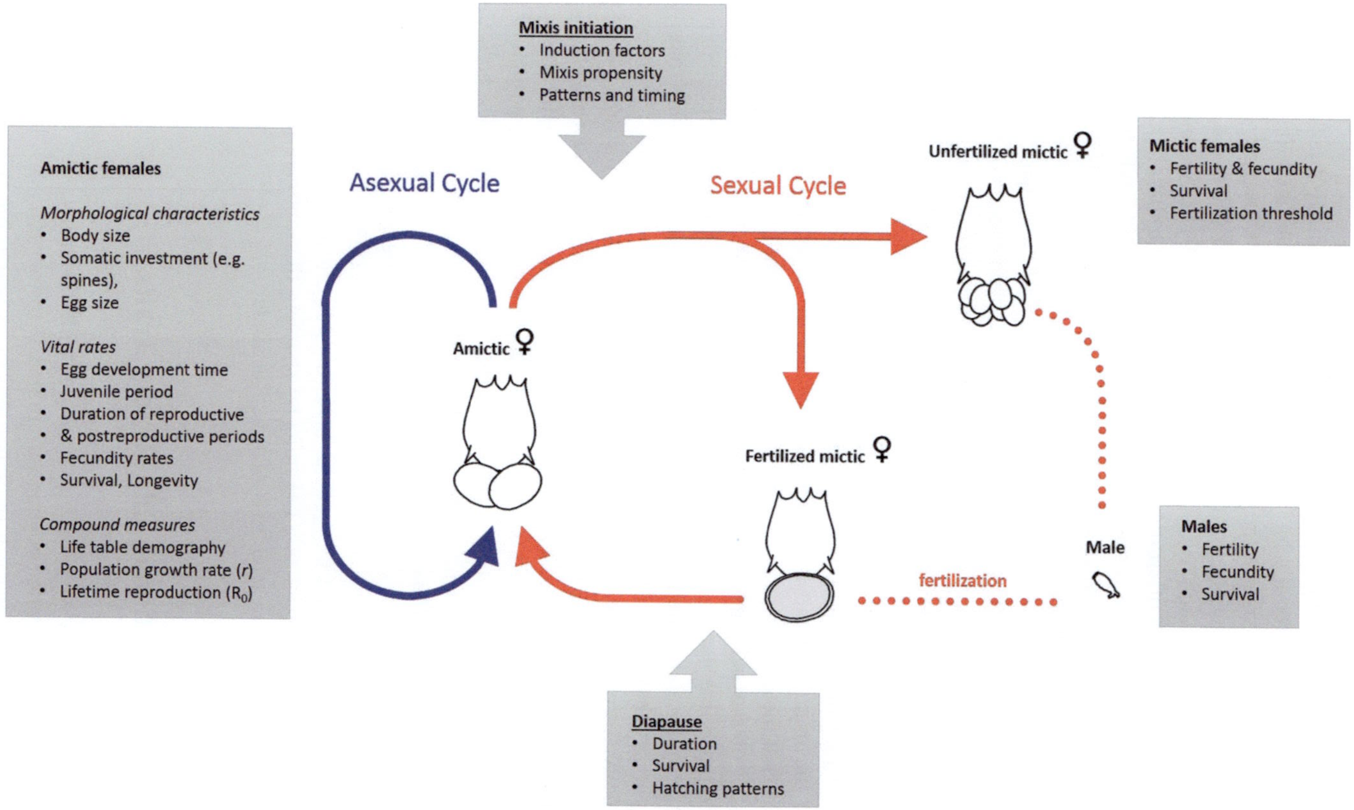

Fig. 7.1 Typical life cycle of a monogonont rotifer. For simplicity, only adult females (no juveniles) are shown. Monogonont rotifers can reproduce either asexually by ameiotic parthenogenesis (*blue*) or sexually (*red*). Asexual females are also called amictic females, and they produce exact genetic copies of themselves (barring mutations). Sexual reproduction is initiated by the production mictic (= sexual) daughters, which produce either haploid males (if not fertilized) or diploid resting eggs (if fertilized). The cycle is completed once amictic females hatch from the resting eggs. Life history traits of the various stages are listed next to the life cycle scheme

crowding chemicals, since in the last decade by far the most work has been conducted on species with density-dependent mixis induction.

Mixis propensity can be defined in two contexts: First, the density threshold for mixis is the population density at which sexual reproduction is initiated. In experimental studies this is often measured as the density at which the first male is observed (in a population growing from low to high density). For a detailed discussion of the methodology of such assays, see Carmona et al. (2011). Second is the ratio of mictic offspring in all offspring of females that received a sufficient crowding stimulus. In many cases this resembles the maximum mixis ratio, which is variable from clone to clone but usually below 100% (e.g., Gilbert 2004). Density thresholds for mixis can be most accurately measured if one or very few asexual individuals are introduced into a relatively large volume (>20 ml), which ensures the population can start to grow from below-threshold concentrations of the mixis factor. By contrast, "maximum mixis ratios" are best estimated if individuals are cultured in small volumes (≤1 ml) with infrequent medium exchange and/or by using conditioned medium from very dense cultures (>50 individuals per ml), which ensures high concentrations of the mixis factor.

The basics of the monogonont life cycle, as depicted in Fig. 7.1, have been established since the middle of the last century. However in the recent years, it has become clear that the entire life cycle may be subject to evolutionary change such that the sexual or asexual parts may be abandoned. For instance, Schröder et al. (2007) found that some populations of the rotifer *Hexarthra* sp. in the Chihuahuan Desert have partially abandoned the asexual part of their life cycle. They found that sexual females in these populations, in contrast to other monogonont rotifers, hatched directly from the resting eggs, and they could be instantly fertilized by males and lay resting eggs. This change in the life cycle is interpreted as an adaptation to the unpredictable and often extremely short growing season in these habitats (Schröder et al. 2007). By contrast, several studies have shown that rotifer populations subjected to high population growth and benign conditions for extended periods of time may completely abandon the sexual part of the life cycle (Bennett and Boraas 1988; Boraas 1983; Fussmann et al. 2003; Stelzer 2008). These obligate parthenogens became unresponsive to the mixis-inducing protein, the crowding stimulus which normally induces sex in this genus (Stelzer 2008). In a later study, Stelzer et al. (2010) established the genetic basis of obligate parthenogenesis (or zero mixis propensity, if viewed as a LH trait) by showing that obligate parthenogens carry two copies of a recessive loss-of-function allele.

7.2 Sources of Variation in Life History Traits

There are two main reasons why life history traits might differ between two individuals: phenotypic plasticity and genetic variation. The former refers to changes in the phenotype due to differences in the environment in which the individual has grown up. The latter refers to differences that are due to the genetic makeup of individuals. Such genetic differences might be fixed within a species, or they may show some degree of variation within or among populations. Of course, a combination of both sources of variation is possible, such that two genotypes might differ in their phenotypic response to environmental factors.

Aside these two sources of variability, individuals of the same clone might differ in their life history traits owing to factors that are not (or at least not directly) dependent on the environment. This includes maternal effects and transgenerational effects (which last for several generations). Finally there may be some inherent developmental stochasticity causing enhanced variability in certain life history traits (e.g., entry and exit of diapause). Such cases are often interpreted as diversified bet-hedging strategies, i.e., the production of variable offspring in anticipation of unpredictable environmental conditions (Schröder 2005).

7.2.1 Phenotypic Plasticity

Studies on the phenotypic plasticity of life history traits have been very popular with rotifers, since clonal reproduction allows the same genotype to be replicated over an essentially unlimited number of factors and factor combinations. In most cases such studies are confined to the asexual phase of the rotifer life cycle, and typically they encompass cohort life tables. Some of these factors have been summarized in Table 7.1, together with commonly observed effects and a few example studies.

7.2.1.1 Environmental Variables

By far the most extensively studied environmental factor is *temperature*. Temperature has several interesting effects on the life history traits in rotifers. First, and like in other ectothermic organisms, the vital rates of rotifers are sped up by increasing temperatures (over a certain temperature range and provided that food is abundant). That is, embryonic development and juvenile development are completed within shorter time periods, individual offspring are produced in shorter time intervals, and individuals have a shorter life expectancy (Halbach 1970; Herzig 1983; Miracle and Serra 1989), although there may be strain-specific differences in these responses (e.g., Xi et al. 2005). Second, if the population growth rate r is considered over a wide temperature range, rotifers show the typical thermal performance curves known from other ectothermic organisms (Huey and Kingsolver 1989), with a single maximum and an asymmetric skew toward lower temperatures (Walczyńska and Serra 2014a). The position of peak is an indication of the temperature to which a species is adapted to, while the breath of the peak performance indicates the degree of specialization, i.e., whether a species is stenotherm or eurytherm. Third, at least some rotifer strains or species obey a pattern termed the "temperature-size rule" (Atkinson 1994), i.e., they mature at a larger body size if cultured at low temperatures (Stelzer 2002; Sun and Niu 2012). According to Stelzer (2002), this is caused

Table 7.1 Life history variation due to phenotypic plasticity. Studies in this category typically use one (or few) clones and manipulate different environmental factors or combinations thereof

Factor	Commonly observed patterns	Example studies
Temperature	Faster embryonic and juvenile development, higher fecundity, and decreased longevity at high temperatures (metabolism acceleration)	Halbach (1970) and Miracle and Serra (1989)
	Population growth rates increase with temperature over a large range, but decline sharply after optimum is exceeded	Walczyńska and Serra (2014a)
	Increase in body size/egg size at low temperatures (in some species)	Stelzer (2002) and Sun and Niu (2012)
Food		
Food type	Some alga species have higher nutritional value than others and yield higher fecundity (mx) and R_0	Flores-Burgos et al. (2005), Geng and Xie (2008), Korstad et al. (1989), and Suchar and Chigbu (2006)
Elemental composition	N and P limitation in the food algae can reduce fecundity and population growth. Rotifers (*Brachionus*) may be more affected by N limitation than P limitation than other aquatic organisms	Hessen et al. (2007), Rothhaupt (1995), and Štrojsová et al. (2009)
Food concentration	Life history variables (r, R_0, mx) usually increase with food concentration, but sometimes reduced longevity	Decline of r at very high food concentrations: Stemberger and Gilbert (1985)
Dietary restriction and intermittent starvation	In some species: increased longevity	Gribble and Welch (2013), Weithoff (2007), and Yoshinaga et al. (1999)
Salinity and pH	Highest population growth rate within an "optimal" salinity or pH range; decrease of fitness in extremely low/high ranges, but at high food levels may compensate and allow a better tolerance to extreme values (–> "broader" peaks)	Miracle and Serra (1989), Oltra and Todolf (1997), Snell (1986), Yin and Niu (2008), and Weisse et al. (2013)
	Increased egg mortality at low pH	
Allelochemicals		
Crowding	Production of mictic daughters	Gilbert (1963) and Stelzer and Snell (2003)
	Faster development, earlier reproduction, decreased longevity in the presence of crowding chemicals; R_0 equal or lower at high densities	Yoshinaga et al. (1999) and Carmona et al. (1994)

(continued)

Table 7.1 (continued)

Factor	Commonly observed patterns	Example studies
Predator	Increased somatic growth (e.g., longer spines), larger body size and egg size; sometimes increase in reproductive rates – at the cost of longevity (Garcia et al. 2007) sometimes decrease in reproductive rates (Stemberger 1988)	Stemberger (1988), Pavón-Meza et al. (2007), and Garcia e al. (2007)
Competitor	Mixed results: *Daphnia* chemicals induced lower fecundity in one study, but another study found that they stimulate growth	Conde-Porcuna (1998) and Guo et al. (2011)
Natural toxins	Lower survival and reproduction at high doses, but species-specific sensitivity to toxins from cyanobacteria	Geng and Xie (2008) and Huang et al. (2012)
Anthropogenic factors		
Toxins (heavy metals, pesticides, etc.)	Lower survival and reproduction at high doses	Guo et al. (2012), Huang et al. (2007), Marcial and Hagiwara (2007), Marcial et al. (2005), Ríos-Arana et al. (2007), and Sarma et al. (2006, 2008)
Nanoparticles	37 nm particles can negatively affect population growth	Snell and Hicks (2011)
Antiaging supplements (e.g., antioxidants)	Increased longevity, in some cases "at the cost of" reduced fecundity	Enesco and Verdone-Smith (1980), Snare et al. (2013), and Snell et al. (2012, 2014)
Endocrine disruptors (pharmaceuticals, hormones, etc.)	Multiple effects, depending on the signaling pathway an endocrine disruptor interferes with, e.g., progesterone: increased production of mictic daughters/resting eggs	Stout et al. (2010)

by a combination of two mechanisms: First, females allocate more resources to individual offspring (i.e., they lay larger eggs), and, second, individuals grow to a larger body size when cultured at low temperatures during their juvenile phase. In *Synchaeta pectinata*, Stelzer (2002) found no evidence supporting that large offspring were growing better at low temperatures, even though small offspring were growing significantly faster at high temperatures. Sun and Niu (2012) showed that offspring from large eggs in *Brachionus calyciflorus*, which were produced at low temperatures, were more starvation-resistant than offspring from small eggs at all temperatures. Furthermore offspring from larger eggs had a higher population growth rate at low temperatures (due to an earlier maturation of large offspring at low temperatures). These observations suggest that the observed temperature-size patterns might be adaptive.

Food quality and quantity are factors often manipulated in life table experiments. In most cases, the speed of juvenile development and early adult fecundity increase with food concentration although at extremely high food concentrations can sometimes diminish fecundity or population growth (Stemberger and Gilbert 1985). Interestingly, a number of studies have shown that longevity may actually increase, if individuals are kept at low food concentrations or if they are subjected to intermittent starvation. This life-extending effect ("dietary restriction") is in line with observations in many other organisms (see this volume; chapter by Kaneko and Yoshinaga). However, not all rotifer species respond to dietary restriction with extended longevity. For example, Weithoff (2007) investigated the effects of dietary restriction on longevity in two rotifer species and found that only *Elosa worallii* responded with an increase in lifespan at the expense of reproduction, while the other rotifer, *Cephalodella* sp., showed both reduced fecundity and survival under dietary restriction.

Mixis rates are also known to be affected by food concentration and food quality. However, unlike in other heterogonic organisms where food scarcity promotes sexual reproduction (e.g., Daphnia, Kleiven et al. 1992), low food concentrations in rotifers seem to suppress the production of offspring (even in the presence of crowding chemicals). Snell (1986) showed that asexual reproduction in *Brachionus* is possible at food concentrations much lower than those required for sexual reproduction. Similar results were found for other environmental variables (temperature, salinity), where environmental extremes reduced sexual reproduction to a greater extent than asexual reproduction (Lubzens et al. 1985; Snell 1986). Several other studies also support the view that low food concentrations suppress, rather than enhance, sexual reproduction in rotifers (summarized in Schröder 2005). Lubzens and Minkoff (1988) reported an effect of maternal diet (aged vs. fresh food algae) that could influence the (maternal) age-dependent pattern of mictic offspring production in *B. plicatilis*: When fed with fresh food algae, the highest proportion of mictic offspring was produced in the middle of the mother's reproductive phase. In contrast, when fed with aged food algae, the highest proportion of mictic offspring was produced at the end of the mother's reproductive phase.

Salinity and *pH* are also important environmental factors and have been used as treatments in many life table studies, often in combination with other factors, such as food concentration or temperature (e.g., Miracle and Serra 1989; Oltra and Todolf 1997; Snell 1986; Weisse et al. 2013). In those cases where a relatively broad range of salinities or pH was considered, life history traits (e.g., fecundity) show a single maximum over the range of the respective factor levels, which indicates that there may be some species-specific or strain-specific optimal pH or salinity level. Interestingly, however, the breadth of this peak may vary with other factors. For example, benign conditions such as high food levels or optimal temperatures led to a broader pH peak in *Cephalodella acidophila* (Weisse et al. 2013). The broader peak in the performance along the pH gradient reflected the "fundamental niche" of this species, whereas the narrower peak, obtained under food limitation and suboptimal temperatures, likely reflects the "realized niche" of this species under more natural conditions. In several rotifer species, a low pH value has been shown to cause increased egg mortality (Weisse et al. 2013; Yin and Niu 2008), which suggests that embryonic development is particularly sensitive to pH stress.

7.2.1.2 Allelochemicals

There are a large number of allelochemicals that can affect life history traits in rotifers. These include chemicals exuded by predators, competitors, or chemicals released by other conspecifics of the same population. The experimental protocol for testing such chemicals is simple: Authors have either used conditioned medium, or they separated "inducers" and "receivers" spatially using 30–50 μm mesh nylon nets, or by culturing individuals in experimental vessels of different size ("self-crowding").

Predator chemicals often elicit life history changes (e.g., body size and shape) that may be interpreted as a defense mechanism. For example, chemicals exuded by *Asplanchna* induce elongated spines in several *Brachionus* and *Keratella* species (Gilbert and Waage 1967; Stemberger and Gilbert 1984, 1987). This enhanced investment into somatic growth may lead to a measurable cost in terms of reduced fecundity, measured as R_0 (Stemberger 1988), which is indicative of a physiological trade-off between somatic growth (ultimately higher survival, in the presence of predators) and reproduction. Many authors, however, could not experimentally confirm such reproductive costs of induced defense structures (reviewed by Gilbert 2013). Moreover, some authors found that reproductive rates may actually be stimulated by the presence of predator chemicals. For example, Garcia et al. (2007) showed that water conditioned by the two predators *Ambystoma mexicanum* and *Acanthocyclops robustus* led to the following life history alterations in *Brachionus havanaensis*: larger body and spine length, earlier reproduction, faster reproduction (but shorter reproductive period), and decreased longevity. Overall population growth (measured as r) was significantly higher in predator-conditioned medium compared to unconditioned medium. The authors concluded that in addition to morphological defenses, *B. havanaensis* seems to compensate losses through predation with enhanced offspring production – at the cost of a shorter lifespan.

Chemical signals released by competitors are another way of competitive interaction, in addition to exploitative competition and (physical) interference competition. Life history changes induced by chemicals of a different species have been investigated by Conde-Porcuna (1998). The author found that *Daphnia*-conditioned water elicited the following life history changes in *Keratella cochlearis*: R_0 and r were decreased – while lifespan, embryonic development time, and juvenile period were not affected. The induced life history changes were probably not simply due to ammonium toxicity, caused by NH_4^+ excretion of *Daphnia*, since he showed that ammonium levels in this experiment were lower than those occurring in natural habitats. In contrast to Conde-Porcuna, Guo et al. (2011) found that population growth of *Brachionus calyciflorus* was stimulated in the presence of chemicals released by its competitor *Daphnia similis*.

Crowding chemicals are an important means of communication in rotifer populations and a source of phenotypic variation for many life history traits. Most notably, crowding chemicals are the trigger for induction of the sexual cycle in many monogonont rotifer species; thus they determine how much a clone invests into asexual and sexual reproduction. The shape of this phenotypic response (i.e., the relationship between the concentration of crowding chemical and % sexual offspring) is likely a threshold function with saturation at a maximum percent of mictic offspring (Serra et al. 2011).

Crowding chemicals might also affect the life history characteristics of asexual females. For example, Carmona et al. (1994) found that clones cultured in small volumes (0.2 ml as opposed to 1 ml) exhibited a lower reproductive output, R_0, and lifespan. At first sight, one might argue that this would be explained by a higher resource depletion in the small-volume group, but the authors note that algae populations in their experiment were growing continuously throughout the experiment. A study by Yoshinaga (1999) avoided such complications by using conditioned medium, which was applied at the same volume in control and crowding treatment. In contrast to Carmona et al. (1994), he found R_0 to be unaltered. However, crowding chemicals appeared to speed up the life cycle, such that animals in conditioned media completed their maturation at smaller body sizes and showed higher intrinsic rates of natural increase (r). They exhibited an earlier fecundity schedule (mx), with a lifespan shorter than that of the control animals.

7.2.1.3 Toxins and Endocrine Disruptors

There are a large number of chemicals, often of anthropogenic origin, with known phenotypic effects on rotifer life history traits. *Toxins* are chemicals that have unambiguously negative effects on affected life history variables. Studies on chronic toxicity usually involve cohort life tables of asexually reproducing females. Alternative experimental designs, which include also sexual stages as well, have been suggested (Preston and Snell 2001), but these seem to be constricted to *Brachionus* spp. as test organism. Heavy metals or pesticides are the common treatments in toxicological studies: Once a certain threshold concentration (no observed effect concentration, NOEC) is exceeded, they may slow development, lower fecundity, or reduce longevity. A detailed treatment of the effects of different classes of toxins in life history variables is beyond the scope of this chapter; interested readers are referred to earlier toxicology reviews (Dahms et al. 2011; Snell and Janssen 1996). *Endocrine disruptors* are substances that specifically interfere with the hormonal system. As such, they may modify allocation patterns and activate physiological trade-offs in an individual, which can theoretically result in enhancement of some life history variables (e.g., García-García et al. 2014; Snell and DesRosiers 2008). Studies have also demonstrated that sexual processes (mixis induction, fertilization, resting egg production) are more sensitive to endocrine disruption, i.e., NOECs are one to two magnitudes lower than those of asexual life history traits (fecundity, survival, R_0, r) (Preston et al. 2000; Snell and Joaquim-Justo 2007).

Over the last two decades, there has been increasing interest in a particular class of endocrine disruptors: vertebrate and invertebrate *hormones* – in particular steroids. As one of the earliest studies in this field, Gallardo et al. (1997) investigated the effect of several vertebrate and invertebrate hormones on *B. plicatilis*: growth hormone (GH), human chorionic gonadotropin (HCG), juvenile hormone (JH), gamma-aminobutyric acid (GABA), β-estradiol (E_2), 20-hydroxyecdysone (HE), triiodothyronine (T_3), and 5-hydroxytryptamine (5-HT). Interestingly, additions of these chemicals led to increases in asexual population growth (GABA, JH, HCG), increased mictic female production (JH, 5-HT, E_2, GABA, HE), or larger body size (JH, GH), which suggested that rotifers use similar components in their hormonal signaling system as do many other invertebrates and vertebrates. Snell and Joaquim-Justo (2007) pointed out that rotifers seem to be particularly sensitive to androgenic (e.g., testosterone) and antiandrogenic substances (e.g., flutamide). Strong proof for a mechanistic basis of endocrine disruption in rotifers came with the discovery of a progesterone receptor in *Brachionus manjavacas* (Stout et al. 2010). The receptor was predominantly located in reproductive organs, including the ovaries, vitellarium, oviduct, and egg in females and the seminal vesicle, rudimentary gut, and sperm duct in males. Females transfected with a double-stranded RNA of a fragment of the receptor (a treatment which resembles a knockdown of the receptor) experienced a 64% reduction of sexual offspring (Stout et al. 2010). This is in line with observations that progesterone addition elicits higher diapausing egg production in this species (Snell and DesRosiers 2008).

7.2.1.4 Other Chemical Substances

There are a few other substances with known effects on rotifer life history variables. For example, a number of *antiaging supplements* have been tested in rotifers, e.g., antioxidants (Bozovic and Enesco 1986; Enesco and Verdone-Smith 1980; Snell et al. 2012). Enesco and Verdone-Smith (1980) obtained a 10% extension in lifespan using alpha-tocopherol (vitamin E) with no adverse effects on fecundity. Snell et al. (2012) tested various antioxidants in isolation and combination and noted that combinations of trolox, N-acetylcysteine, L-carnosine, and EUK-8 were particularly effective and could extend lifespan up to 20%. Other examples for dietary supplements with life-extending capabilities include red algae extracts (Snare et al. 2013) or small molecules such as rapamycin and JNK inhibitor, two chemicals interfering the TOR kinase pathway and Jun-N-terminal kinase (JNK) signaling pathway, which are two phylogenetically conserved pathways involved in energy allocation and stress response (Snell et al. 2014). It is interesting to note that these two chemicals did not only increase lifespan but also resulted in alteration of the fecundity pattern: Exposure to 1 μM rapamycin resulted in lower reproduction in the early adult life but a longer reproductive period. Exposure to 1 μM JNK inhibitor substantially reduced fecundity at all ages; in fact, the eggs that were produced under JNK inhibitor exposure did not hatch (Snell et al. 2014). These observations suggest that the two chemicals target the underlying mechanisms of major life history trade-offs (current vs. future reproduction, reproduction vs. survival).

Table 7.2 Life history variation due to genetic differences. Studies in this category typically use many different clones or populations with mixed-clonal composition

Source of variation	Experimental approach	Example studies	Life history traits
Among species	Comparative approach: assess covariation of two or more traits among *n* members of a taxonomic group cultured under identical conditions (=common garden)	Stemberger and Gilbert (1985) and Walz (1995)	Population growth, egg size (correlation with body size)
		Walczyńska and Serra (2014b)	Hatching pattern (correlation with body size)
	Comparative approach (phylogenetically controlled)	Stelzer et al. (2011)	Body size, egg size (correlation with genome size)
Among populations	Sample multiple clones from two or more populations and quantify among-population variation in a common garden		
	Local adaptation: + use multiple types of common gardens, each exhibiting the habitat conditions of one of the source populations + estimate magnitude of among-population variation expected under neutral evolution	Alcantara-Rodriguez et al. (2012) and Campillo et al. (2009)	Fecundity, population growth, mixis propensity
Within populations	Quantify trait variation among clones of the same population under identical conditions (common garden)	Snell and King (1977)	Lifespan, fecundity
		Becks and Agrawal (2010)	Mixis propensity
	Assess a trait's change in response to selection (in experimental or wild populations). The source population is composed of multiple clones	Becks and Agrawal (2010), Carmona et al. (2009), Fussmann et al. (2003), and Smith and Snell (2012)	Mixis propensity
		Smith and Snell (2014)	Lifespan, fecundity

7.2.2 Genetic Variation

There are a substantial number of studies on genetic variation of life history traits in rotifers. Most of them compare life history traits of different rotifer clones under identical conditions (i.e., common garden experiments; see Table 7.2). One-way analysis of variance can then be used to identify significant variation among clones within a population. Alternatively, nested ANOVA can be used for testing whether populations differ genetically from each other (i.e., factor "clone" nested within factor "population").

However there are two possible pitfalls with this experimental design. First, it requires that maternal and grandmaternal environments do not contribute to variation of the experimental group. A precaution against such maternal effects is to culture not only the experimental animals but also their ancestors under the same common garden conditions. Another method consists in splitting up clonal lines into sublines and to partition out variance due to maternal effects with a nested ANOVA (Lynch and Walsh 1998). Second, care must be taken that sampled clones are assigned to the correct species, since an unintentional mix of several cryptic species would inevitably confound estimates of population-level variation. Therefore, some rotifer researchers use DNA barcoding in addition to morphological species assignment to ensure the correct species status (e.g., Campillo et al. 2009; Smith and Snell 2012).

7.2.2.1 Within-Population Variation

Several studies have shown that rotifer populations may harbor significant amounts of heritable variation in the trait "mixis propensity" (see Table 7.2). There are two experimental approaches for measuring within-population variation of life history traits in rotifers. The first involves variance partitioning and testing for significant among-clone variation for a certain life history trait (e.g., Aparici et al. 2001; Becks and Agrawal 2010). The second approach is more indirect and involves selection applied to a base population composed of several clones. If the population is variable with respect to the life history trait of interest, its composition will change over time, and selectively favored variants will dominate future generations. For example, Smith and Snell (2012) used 15 *Brachionus plicatilis* clones, exposed mixed populations in chemostats, and followed an experimental protocol that was either mimicking ephemeral or permanently filled aquatic habitats. Smith and Snell (2012) found that ephemeral populations evolved higher mixis propensity compared to permanent populations. As another example, Carmona et al. (2009) studied evolutionary changes in mixis propensity in a field population of *B. plicatilis* during the growing season. They found that the frequency of clones with low propensity for sex increased during the growing season. This pattern was likely caused by selection for low mixis propensity in a well-adapted population under benign conditions, and it fits well to observations of selection against sex in benign lab environments (Stelzer 2011).

In contrast, Becks and Agrawal (2010, 2012) studied the evolution of mixis rates in genetically variable populations where not all individuals are optimally adapted. This was either due to frequent migration between two contrasting habitats (Becks and Agrawal 2010) or because individuals just colonized an unfamiliar habitat (Becks and Agrawal 2012). The authors showed that high mixis rates can evolve in both conditions. This was likely due to the mechanism that clones with high mixis were more likely to generate variable offspring that could better deal with the changed habitat conditions. Thus, in both cases, higher mixis rates are thought to yield indirect benefits, because sex would allow the generation of better adapted genotypes.

7.2.2.2 Between-Populations Variation

Only few studies have rigorously examined between-population variation of life history traits in rotifers. This is probably due to the high number of replicates required for a variance-partitioning approach (i.e., replicate nested within clone and clone nested within population). One of the few exceptions is the study of Campillo et al. (2009) who measured 30 *B. plicatilis* clones from each of 6 geographic populations with a total of 16 replicates per clone for 6 life history traits, e.g., female fecundity and production rates of sexual offspring (males, diapausing eggs). Between populations variation may be selectively neutral or it may be adaptive ("local adaptation"). One way to test if differences among populations are indeed selectively favored is to calculate F_{ST} values and Q_{ST} values for pairs of populations. F_{ST} refers to neutral genetic differentiation and can be estimated using molecular markers such as microsatellites, while Q_{ST} refers phenotypical differentiation in a certain quantitative trait (e.g., reproductive output). Campillo et al. (2009) found that $Q_{ST} > F_{ST}$ for traits related to sexual reproduction, which is indicative of divergent selection and could have been caused by differences among the habitats with regard to length of growing season or differences in habitat unpredictability. Another experimental design used in testing for local adaptation involves culturing individuals in multiple environments (local vs. foreign). If a population is locally adapted, it should perform best in its own, local environment (compared to other environments), and it should be the best performer in its local environment (compared to populations sampled from other environments).

7.2.2.3 Among-Species Variation

For species comparisons the abovementioned variance-partitioning approach becomes impractical, since it is impossible to include and randomize a large number of clones and populations. An alternative consists in the "comparative method," i.e., sampling only one or few clones per species and using correlations among several life history variables in a large number of species within a certain taxonomic group of rotifers (e.g., a genus) or of rotifers as whole. Such studies have shown, for example, a positive correlation between rotifer body weight and *r* (Stemberger and Gilbert 1985) or a negative relationship between body volume and relative egg volume (Walz 1995). The latter indicates that small rotifers invest proportionally more resources into individual offspring than large rotifers. However, the comparative method has its own caveats and limitations. First, since it relies on the calculation of correlation coefficients, it should include a sufficient number of species. Thus, two species comparisons are not considered useful in this regard (Garland and Adolph 1994). Second, some correction for phylogenetic nonindependence may be necessary, especially if a narrow taxonomic group is considered (Felsenstein 1985; Garland et al. 1992). For example, Stelzer et al. (2011) showed that genome size in the *Brachionus plicatilis* complex is significantly correlated to egg size and body size, even though the latter became nonsignificant after controlling for phylogenetic nonindependence. In most cases such a correction will give a more conservative estimate of the correlation coefficient, because it puts less weight on species that are very closely related.

7.2.3 Other (Nongenetic) Sources of Variation

There is evidence that *maternal effects* can be an important source of variation in life history traits of rotifers. For example, Schröder and Gilbert (2009) showed in some clones of *Brachionus calyciflorus* that spine length and body size in offspring increase with birth order (maternal age at egg production). Interestingly, spine lengths in their study increased to lengths formerly observed only in the presence of *Asplanchna*. A study of Hagiwara et al. (2005) provides an example of a maternal effect that influences the production of mictic offspring: If stem females received a short starvation treatment during their juvenile period, their daughters exhibited a higher rate of mictic offspring production. This effect, however, was only present in stem females that had directly hatched from resting eggs and was not found in hatchlings from asexual eggs (Hagiwara et al. 2005).

Maternal effects are not always as clear-cut as in the above examples. In fact, the distinction between maternal effects and phenotypic plasticity becomes blurred if the developmental switch that results in phenotypic changes is activated at a time when the focal individual is still an egg inside the mother's body. For example, it is difficult to tell whether the crowding factor that induces mixis acts directly on the developing embryo or if the factor acts indirectly on maternal physiology, such as by causing the brain to secrete factor affecting the oocyte (Gilbert 2003). An experiment by Gilbert (2007) employed an experimental protocol in which amictic females were transferred from low to high population densities or vice versa in the middle of their adult lifespan. The corresponding response in the change in mixis happened without significant time lags – this suggests that whether an individual becomes mictic is determined during oogenesis and before oviposition (Gilbert 2007).

In contrast to maternal effects, *transgenerational effects* are phenotypically visible over much longer periods than one generation. For example, Gilbert (2002) showed in hatchlings from diapausing eggs of *Brachionus calyciflorus* that mixis propensity increases gradually from zero to a maximum value (~50%) within the first 12–18 asexual generations and stays constant after that. This result was later corroborated for more species and strains, and this study also revealed that there is some clonal variation for this delayed mixis response to crowding (Schröder and Gilbert 2004).

The mechanisms responsible for these "other (nongenetic) sources of variation" are not fully understood to date. Maternal effects might be explained simply by cytoplasmatic factors that are produced in the mother's vitellarium and transferred to the developing oocyte, where they subsequently alter gene and tissue expression during the embryonal phase. Alternatively, in the case of maternal effect on egg size or body size of offspring, mothers might provide simply more/less cytoplasm to individual eggs during oogenesis. Such mechanisms are not very plausible in the case of transgenerational effects. Possibly, the latter might be caused by epigenetic changes at a certain stage (e.g., resting egg formation) that are gradually and stochastically lost during consecutive asexual generations. The elucidation of such mechanisms remains a challenge for future studies.

7.3 Conclusions and Future Perspectives

Life history theory provides a general framework of how key events and processes through an individual lifetime (age at reproduction, reproductive investment, survival, etc.) are optimized by evolution under a given set of internal trade-offs and other constraints. Many studies on life history variation in rotifers apply these concepts and try to predict (or test) which life history variants should evolve under given biotic or abiotic conditions. Such studies have shown, for example, how mixis rates evolve in benign vs. unpredictable environments. Another set of studies uses specific life history traits as endpoint (e.g., R_0 or r), while the actual subject of interest is the "dependent variable," such as environmental variables (characterization of the autecological niche) or toxins (environmental toxicology). Finally some research fields focus on specific life history variables (like mortality rates or longevity; aging research). In conclusion, life history studies conducted with rotifers are a blend of quite distinct research fields, each with a different motivation, focus, or conceptual background.

Nevertheless, the basic principles of life history theory can be identified in all of these research fields. For example, trade-offs between survival and reproduction are apparent in many cases of chemicals targeting signaling pathways in rotifers (e.g., endocrine disruptors, antiaging supplements). Such signaling interconnections are promising in two directions: First, a targeted manipulation with certain chemicals (e.g., progesterone) may allow to identify the basic cellular and molecular mechanism underlying a life history trait. Knowledge of such mechanisms is of key importance in a better understanding of genetic variation in life history traits – within and between populations. Such research should also extend to nongenetic mechanisms, for example, maternal effects and epigenetic mechanisms. Indeed, rotifers are excellent model organisms for studying these more "exotic" mechanisms due to the ease of replicating genotypes by clonal reproduction. Second, the application of general life history theory may provide additional insights. For example, studies on aging might benefit from including other fitness components (e.g., juvenile development, fecundity) or fitness indicators (R_0, r, resting egg production). Thus the identification of the mechanistic basis of key life history events, together with a strong integration of the general concepts of life history theory, might be a promising agenda for future studies on rotifer life history evolution.

References

Alcantara-Rodriguez JA, Ciros-Perez J, Ortega-Mayagoitia E, Serrania-Soto CR, Piedra-Ibarra E (2012) Local adaptation in populations of a *Brachionus* group *plicatilis* cryptic species inhabiting three deep crater lakes in Central Mexico. Freshw Biol 57:728–740. doi:10.1111/j.1365-2427.2012.02738.x

Aparici E, Carmona MJ, Serra M (1998) Sex allocation in haplodiploid cyclical parthenogens with density-dependent proportion of males. Am Nat 152:652–657. doi:10.1086/286196

Aparici E, Carmona M, Serra M (2001) Variability for mixis initiation in *Brachionus plicatilis*. Hydrobiologia 446(447):45–50

Atkinson D (1994) Temperature and organism size—a biological law for ectotherms? Adv Ecol Res 25:1–58

Becks L, Agrawal AF (2010) Higher rates of sex evolve in spatially heterogeneous environments. Nature 468:89–92. doi:10.1038/nature09449

Becks L, Agrawal AF (2012) The evolution of sex is favoured during adaptation to new environments. PLoS Biol 10:11. doi:10.1371/journal.pbio.1001317

Bennett WN, Boraas ME (1988) Isolation of a fast-growing strain of the rotifer *Brachionus calyciflorus* Pallas using turbidostat culture. Aquaculture 73:27–36

Boraas ME (1983) Population dynamics of food-limited rotifers in two-stage chemostat culture. Limnol Oceanogr 28:546–563

Bozovic V, Enesco HE (1986) Effect of antioxidants on rotifer lifespan and activity. Age 9:41–45

Campillo S, Garcia-Roger EM, Carmona MJ, Gomez A, Serra M (2009) Selection on life-history traits and genetic population divergence in rotifers. J Evol Biol 22:2542–2553

Carmona MJ, Serra M, Miracle MR (1994) Effect of population density and genotype on life-history traits in the rotifer *Brachionus plicatilis* O.F. Müller. J Exp Mar Biol Ecol 182:223–235

Carmona MJ, Dimas-Flores N, Garcia-Roger EM, Serra M (2009) Selection of low investment in sex in a cyclically parthenogenetic rotifer. J Evol Biol 22:1975–1983. doi:10.1111/j.1420-9101.2009.01811.x

Carmona M, Dimas-Flores N, Montero-Pau J, Serra M (2011) Effect of experimental methodology on estimation of density at sex initiation in cyclically parthenogenetic rotifers. Hydrobiologia 662:131–139

Chen F, Xi Y-L, Ge Y-L, Xu X-P (2005) Life history traits of mictic females in two strains of *Brachionus calyciflorus* (Rotifera). J Freshw Ecol 20:321–326

Conde-Porcuna J (1998) Chemical interference by *Daphnia* on *Keratella*: a life table experiment. J Plankton Res 20:1637–1644

Dahms HU, Hagiwara A, Lee JS (2011) Ecotoxicology, ecophysiology, and mechanistic studies with rotifers. Aquat Toxicol 101:1–12. doi:10.1016/j.aquatox.2010.09.006

Enesco HE, Verdone-Smith C (1980) Alpha-tocopherol increases lifespan in the rotifer Philodina. Exp Gerontol 15:335–338

Felsenstein J (1985) Phylogenies and the comparative method. Am Nat 125:1–15

Flores-Burgos J, Sarma S, Nandini S (2005) Effect of single species or mixed algal (*Chlorella vulgaris* and *Scenedesmus acutus*) diets on the life table demography of *Brachionus calyciflorus* and *Brachionus patulus* (Rotifera: Brachionidae). Acta Hydrochim Hydrobiol 33:614–621

Fussmann GF, Ellner SP, Hairston NG Jr (2003) Evolution as a critical component of plankton dynamics. P Roy Soc B-Biol Sci 270:1015–1022. doi:10.1098/rspb.2003.2335

Gallardo WG, Hagiwara A, Tomita Y, Soyano K, Snell TW (1997) Effect of some vertebrate and invertebrate hormones on the population growth, mictic female production, and body size of the marine rotifer *Brachionus plicatilis* Müller. Hydrobiologia 358:113–120

Garcia C, De Jesús Chaparro-Herrera D, Nandini S, Sarma S (2007) Life-history strategies of *Brachionus havanaensis* subject to kairomones of vertebrate and invertebrate predators. Chem Ecol 23:303–313

García-García G, Sarma SSS, Núñez-Orti AR, Nandini S (2014) Effects of the mixture of two endocrine disruptors (ethinylestradiol and levonorgestrel) on selected ecological endpoints of *Anuraeopsis fissa* and *Brachionus calyciflorus* (Rotifera). Int Rev Hydrobiol 99:166–172. doi:10.1002/iroh.201301721

Garland T, Adolph SC (1994) Why not to do two-species comparative studies: limitations on inferring adaptation. Physiol Zool:797–828

Garland T, Harvey PH, Ives AR (1992) Procedures for the analysis of comparative data using phylogenetically independent contrasts. Syst Biol 41:18–32

Geng H, Xie P (2008) Experimental studies on the effects of toxic Microcystis aeruginosa PCC7820 on the survival and reproduction of two freshwater rotifers Brachionus calyciflorus and Brachionus rubens. Ecotoxicology 17:709–715

Gilbert JJ (1963) Mictic female production in the rotifer *Brachionus calyciflorus*. J Exp Zool 153:113–124. doi:10.1002/jez.1401530204

Gilbert JJ (2002) Endogenous regulation of environmentally induced sexuality in a rotifer: a multigenerational parental effect induced by fertilisation. Freshw Biol 47:1633–1641. doi:10.1046/j.1365-2427.2002.00900.x

Gilbert JJ (2003) Environmental and endogenous control of sexuality in a rotifer life cycle: developmental and population biology. Evol Dev 5:19–24

Gilbert JJ (2004) Population density, sexual reproduction and diapause in monogonont rotifers: new data for *Brachionus* and a review. J Limnol 63:32–36

Gilbert JJ (2007) Induction of mictic females in the rotifer *Brachionus*: oocytes of amictic females respond individually to population-density signal only during oogenesis shortly before oviposition. Freshw Biol 52:1417–1426

Gilbert JJ (2013) The cost of predator-induced morphological defense in rotifers: experimental studies and synthesis. J Plankton Res 35:461–472. doi:10.1093/plankt/fbt017

Gilbert JJ, Waage JK (1967) *Asplanchna*, *Asplanchna*-substance, and posterolateral spine length variation of the rotifer *Brachionus calyciflorus* in a natural environment. Ecology 14:1027–1031

Gomez A, Serra M (1996) Mate choice in male *Brachionus plicatilis* rotifers. Funct Ecol 10:681–687. doi:10.2307/2390502

Gribble KE, Welch DB (2013) Life-span extension by caloric restriction is determined by type and level of food reduction and by reproductive mode in *Brachionus manjavacas* (Rotifera). J Gerontol A: Biol Sci Med Sci 68:349–358. doi:10.1093/gerona/gls170

Guo RX, Snell TW, Yang JX (2011) Ecological strategy of rotifer (*Brachionus calyciflorus*) exposed to predator- and competitor-conditioned media. Hydrobiologia 658:163–171. doi:10.1007/s10750-010-0459-5

Guo R, Ren X, Ren H (2012) Effects of dimethoate on rotifer *Brachionus calyciflorus* using multigeneration toxicity tests. J Environ Sci Health B 47:883–890

Hagiwara A, Kadota Y, Hino A (2005) Maternal effect by stem females in *Brachionus plicatilis*: effect of starvation on mixis induction in offspring. Hydrobiologia 546:275–279. doi:10.1007/s10750-005-4208-0

Halbach U (1970) Einfluß der Temperatur auf die Populationsdynamik des planktischen Rädertieres B*rachionus calyciflorus*. Oecologia 4:176–207

Herzig A (1983) Comparative studies on the relationship between temperature and duration of embryonic development of rotifers. Hydrobiologia 104:237–246. doi:10.1007/BF00045974

Hessen DO, Jensen TC, Kyle M, Elser JJ (2007) RNA responses to N- and P-limitation; reciprocal regulation of stoichiometry and growth rate in *Brachionus*. Funct Ecol 21:956–962. doi:10.1111/j.1365-2435.2007.01306.x

Huang L, Xi Y-L, Zha C-W, Zhao L-L (2007) Effect of aldrin on life history characteristics of rotifer *Brachionus calyciflorus* Pallas. Bull Environ Contam Toxicol 79:524–528

Huang L, Xi Y, Xu X, Wen X (2012) Responses in population growth and reproduction of the freshwater rotifer *Brachionus calyciflorus* to microcystin-LR at different temperatures. Ann Limnol Int J Limnol 48. Cambridge University Press:383–390

Huey RB, Kingsolver JG (1989) Evolution of thermal sensitivity of ectotherm performance. Trends Ecol Evol 4:131–135. http://dx.doi.org/10.1016/0169-5347(89)90211-5

Kauler P, Enesco HE (2011) The effect of temperature on life history parameters and cost of reproduction in the rotifer *Brachionus calyciflorus*. J Freshw Ecol 26:399–408

Kleiven OT, Larsson P, Hobæk A (1992) Sexual reproduction in *Daphnia magna* requires three stimuli. Oikos 65:197–206. doi:10.2307/3545010

Korstad J, Olsen Y, Vadstein O (1989) Life-history characteristics of *Brachionus plicatilis* (Rotifera) fed different algae. Hydrobiologia 186:43–50. doi:10.1007/Bf00048895

Lubzens E, Minkoff G (1988) Influence of the age of algae fed to rotifers *Brachionus plicatilis* Mueller on the expression of mixis in their progenies. Oecologia 75:430–435

Lubzens E, Minkoff G, Marom S (1985) Salinity dependence of sexual and asexual reproduction in the rotifer *Brachionus plicatilis*. Mar Biol 85:123–126

Lynch M, Walsh B (1998) Genetics and analysis of quantitative traits. Sinauer, Sunderland

Marcial HS, Hagiwara A (2007) Effect of diazinon on life stages and resting egg hatchability of rotifer *Brachionus plicatilis*. Hydrobiologia 593:219–225

Marcial HS, Hagiwara A, Snell TW (2005) Effect of some pesticides on reproduction of rotifer *Brachionus plicatilis* Müller. Hydrobiologia 546:569–575

Miracle MR, Serra M (1989) Salinity and temperature influence in rotifer life history characteristics. Hydrobiologia 186:81–102

Nogrady T, Wallace RL, Snell TW (1993) Rotifera: biology, ecology and systematics. SPB Academic Publishing, The Hague

Oltra R, Todolf R (1997) Effects of temperature, salinity and food level on the life history traits of the marine rotifer *Synchaera cecilia* valentina, n. subsp. J Plankton Res 19:693–702

Preston BL, Snell TW (2001) Full life-cycle toxicity assessment using rotifer resting egg production: implications for ecological risk assessment. Environ Pollut 114:399–406. http://dx.doi.org/10.1016/S0269-7491(00)00232-3

Preston BL, Snell TW, Robertson TL, Dingmann BJ (2000) Use of freshwater rotifer *Brachionus calyciflorus* in screening assay for potential endocrine disruptors. Environ Toxicol Chem 19:2923–2928. doi:10.1897/1551-5028(2000)019<2923:Uofrbc>2.0.Co;2

Ríos-Arana JV, Walsh EJ, Ortiz M (2007) Interaction effects of multi-metal solutions (As, Cr, Cu, Ni, Pb and Zn) on life history traits in the rotifer *Plationus patulus*. J Environ Sci Health A 42:1473–1481

Roff DA (1992) Evolution of life histories: theory and analysis. Springer, New York

Rothhaupt KO (1995) Algal nutrient limitation affects rotifer growth rate but not ingestion rate. Limnol Oceanogr 40:1201–1208

Sarma SSS, Nandini S, Gulati RD (2002) Cost of reproduction in selected species of zooplankton (rotifers and cladocerans). Hydrobiologia 481:89–99. doi:10.1023/A:1021265104165

Sarma S, Martinez-Jeronimo F, Ramirez-Perez T, Nandini S (2006) Effect of cadmium and chromium toxicity on the demography and population growth of *Brachionus calyciflorus* and *Brachionus patulus* (Rotifera). J Environ Sci Health A 41:543–558

Sarma S, Brena-Bustamante P, Nandini S (2008) Body size and population growth of *Brachionus patulus* (Rotifera) in relation to heavy metal (copper and mercury) concentrations. J Environ Sci Health A 43:547–553

Schröder T (2005) Diapause in monogonont rotifers. Hydrobiologia 546:291–306

Schröder T, Gilbert JJ (2004) Transgenerational plasticity for sexual reproduction and diapause in the life cycle of monogonont rotifers: intraclonal, intraspecific and interspecific variation in the response to crowding. Funct Ecol 18:458–466

Schröder T, Gilbert JJ (2009) Maternal age and spine development in the rotifer *Brachionus calyciflorus*: increase of spine length with birth orders. Freshw Biol 54:1054–1065

Schröder T, Howard S, Arroyo ML, Walsh EJ (2007) Sexual reproduction and diapause of *Hexarthra* sp. (Rotifera) in short-lived ponds in the Chihuahuan Desert. Freshw Biol 52:1033–1042. doi:10.1111/j.1365-2427.2007.01751.x

Serra M, King CE (1999) Optimal rates of bisexual reproduction in cyclical parthenogens with density-dependent growth. J Evol Biol 12:263–271

Serra M, Smith HA, Weitz JS, Snell TW (2011) Analysing threshold effects in the sexual dynamics of cyclically parthenogenetic rotifer populations. Hydrobiologia 662:121–130. doi:10.1007/s10750-010-0517-z

Smith HA, Snell TW (2012) Rapid evolution of sex frequency and dormancy as hydroperiod adaptations. J Evol Biol 25:2501–2510. doi:10.1111/j.1420-9101.2012.02614.x

Smith HA, Snell TW (2014) Differential evolution of lifespan and fecundity between asexual and sexual females in a benign environment. Int Rev Hydrobiol 99:117–124

Snare DJ, Fields AM, Snell TW, Kubanek J (2013) Lifespan extension of rotifers by treatment with red algal extracts. Exp Gerontol 48:1420–1427. doi:10.1016/j.exger.2013.09.007

Snell TW (1986) Effect of temperature, salinity and food level on sexual and asexual reproduction in *Brachionus plicatilis*. Mar Biol 92:157–162

Snell TW, DesRosiers NJD (2008) Effect of progesterone on sexual reproduction of *Brachionus manjavacas* (Rotifera). J Exp Mar Biol Ecol 363:104–109. doi:10.1016/j.jembe.2008.06.031

Snell TW, Hicks DG (2011) Assessing toxicity of nanoparticles using *Brachionus manjavacas* (Rotifera). Environ Toxicol 26:146–152. doi:10.1002/tox.20538

Snell TW, Janssen CR (1996) Rotifers in ecotoxicology: a review. Rotifera VII. Springer, Tokyo, pp 231–247

Snell T, Joaquim-Justo C (2007) Workshop on rotifers in ecotoxicology. Hydrobiologia 593:227–232. doi:10.1007/s10750-007-9045-x

Snell TW, King CE (1977) Lifespan and fecundity patterns in rotifers: the cost of reproduction. Evolution 31:882–890

Snell TW et al (2006) A protein signal triggers sexual reproduction in *Brachionus plicatilis* (Rotifera). Mar Biol 149:763–773. doi:10.1007/s00227-006-0251-2

Snell TW, Fields AM, Johnston RK (2012) Antioxidants can extend lifespan of *Brachionus manjavacas* (Rotifera), but only in a few combinations. Biogerontology 13:261–275

Snell TW, Johnston RK, Rabeneck B, Zipperer C, Teat S (2014) Joint inhibition of TOR and JNK pathways interacts to extend the lifespan of *Brachionus manjavacas* (Rotifera). Exp Gerontol 52:55–69. http://dx.doi.org/10.1016/j.exger.2014.01.022

Stearns SC (1992) The evolution of life histories. Oxford University Press, Oxford

Stelzer CP (2001) Resource limitation and reproductive effort in a planktonic rotifer. Ecology 82:2521–2533

Stelzer CP (2002) Phenotypic plasticity of body size at different temperatures in a planktonic rotifer: mechanisms and adaptive significance. Funct Ecol 16:835–841

Stelzer CP (2008) Obligate asex in a rotifer and the role of sexual signals. J Evol Biol 21:287–293. doi:10.1111/j.1420-9101.2007.01437.x

Stelzer CP (2011) The cost of sex and competition between cyclical and obligate parthenogenetic rotifers. Am Nat 177:E43–E53. doi:10.1086/657685

Stelzer CP, Snell TW (2003) Induction of sexual reproduction in *Brachionus plicatilis* (Monogononta, Rotifera) by a density-dependent chemical cue. Limnol Oceanogr 48:939–943

Stelzer CP, Schmidt J, Wiedlroither A, Riss S (2010) Loss of sexual reproduction and dwarfing in a small metazoan. PLoS One 5:e12854. doi:12810.11371/journal.pone.0012854. doi:10.1371/journal.pone.0012854

Stelzer CP, Riss S, Stadler P (2011) Genome size evolution at the speciation level: the cryptic species complex *Brachionus plicatilis* (Rotifera). BMC Evol Biol 11:90

Stemberger RS (1988) Reproductive costs and hydrodynamic benefits of chemically-induced defenses in *Keratella testudo*. Limnol Oceanogr 33:593–606

Stemberger RS, Gilbert JJ (1984) Spine development in the rotifer *Keratella cochlearis* – induction by cyclopoid copepods and *Asplanchna*. Freshw Biol 14:639–647. doi:10.1111/j.1365-2427.1984.tb00183.x

Stemberger RS, Gilbert JJ (1985) Body size, food concentration, and population growth in planktonic rotifers. Ecology 66:1151–1159

Stemberger RS, Gilbert JJ (1987) Multiple-species induction of morphological defenses in the rotifer *Keratella testudo*. Ecology 68:370–378

Stout EP, La Clair JJ, Snell TW, Shearer TL, Kubanek J (2010) Conservation of progesterone hormone function in invertebrate reproduction. Proc Natl Acad Sci U S A 107:11859–11864

Štrojsová M, Suga K, Hagiwara A, Vrba J (2009) Effect of food quantity and quality on population growth rate and digestive activity in the euryhaline rotifer *Brachionus plicatilis* Müller. Int Rev Hydrobiol 94:706–719

Suchar VA, Chigbu P (2006) The effects of algae species and densities on the population growth of the marine rotifer *Colurella dicentra*. J Exp Mar Biol Ecol 337:96–102

Sun D, Niu C (2012) Adaptive significance of temperature-induced egg size plasticity in a planktonic rotifer, *Brachionus calyciflorus*. J Plankton Res. doi:10.1093/plankt/fbs050

Walczyńska A, Serra M (2014a) Inter- and intraspecific relationships between performance and temperature in a cryptic species complex of the rotifer *Brachionus plicatilis*. Hydrobiologia 734:17–26. doi:10.1007/s10750-014-1859-8

Walczyńska A, Serra M (2014b) Species size affects hatching response to different temperature regimes in a rotifer cryptic species complex. Evol Ecol 28:131–140

Walz N (1995) Rotifer populations in plankton communities: energetics and life history strategies. Experientia 51:437–453

Weisse T, Laufenstein N, Weithoff G (2013) Multiple environmental stressors confine the ecological niche of the rotifer *Cephalodella acidophila*. Freshw Biol 58:1008–1015. doi:10.1111/Fwb.12104

Weithoff G (2007) Dietary restriction in two rotifer species: the effect of the length of food deprivation on life span and reproduction. Oecologia 153:303–308

Xi Y-L, Huang X-F, Jin H-J (2001) Life history characteristics of three types of females in *Brachionus calyciflorus* Pallas (Rotifera) fed different algae. Hydrobiologia 446/447:95–98

Xi Y-L, Ge Y-L, Chen F, Wen X-L, Dong L-L (2005) Life history characteristics of three strains of *Brachionus calyciflorus* (Rotifera) at different temperatures. J Freshw Ecol 20:707–713

Yin XW, Niu CJ (2008) Effect of pH on survival, reproduction, egg viability and growth rate of five closely related rotifer species. Aquat Ecol 42:607–616. doi:10.1007/s10452-007-9136-9

Yoshinaga T, Hagiwara A, Tsukamoto K (1999) Effect of conditioned media on the asexual reproduction of the monogonont rotifer *Brachionus plicatilis* O.F. Müller. Hydrobiologia 412:103–110

Chapter 8
Aging and Lifespan in the Rotifer

Gen Kaneko and Tatsuki Yoshinaga

Abstract Rotifers have been used as a model in aging studies due to their several advantages such as short lifespan, ease of culture, and asexual reproduction that enables clonal culture. Recent molecular phylogenetic analyses revealed that rotifers belong to the superphylum Lophotrochozoa and phylogenetically distant from major invertebrate models. Taking advantage of this unique phylogenetic position and development of molecular technologies, rotifers are now recognized to be a traditional but also emerging model in aging studies. This chapter summarizes current knowledge of rotifer aging in association with some traditional aging theories and future prospects. Reevaluation of past works, as well as future advances in rotifer studies, would fill the taxonomic gaps and help us to understand aging mechanisms from a broad perspective.

8.1 Introduction

Many organisms undergo universal and progressive decline of body function with age, which results in increased risk of death in later life. This phenomenon is called "aging" or "senescence," while semantically aging simply means the process of growing old and senescence implies the age-dependent functional impairment. To understand the fundamental mechanism of aging, numerous investigations have been conducted, leading to the establishment of hundreds of aging theories. Recently, however, it is increasingly recognized that aging patterns are unexpectedly diverse; for example, a bdelloid rotifer *Macrotrachela* sp. showed an age-dependent

G. Kaneko (✉)
Department of Aquatic Bioscience, Graduate School of Agricultural and Life Sciences, The University of Tokyo, Tokyo, Japan

Department of Radiology & Biomedical Imaging, School of Medicine, Yale University, New Haven, CT, USA

School of Arts & Sciences, University of Houston–Victoria, Victoria, TX, USA
e-mail: KanekoG@uhv.edu

T. Yoshinaga
School of Marine Biosciences, Kitasato University, Tokyo, Japan

A. Hagiwara, T. Yoshinaga (eds.), *Rotifers*, Fisheries Science Series,
DOI 10.1007/978-981-10-5635-2_8

increase in mortality like most mammals, whereas the hydra *Hydra magnipapillata* hardly increased mortality with age and was estimated to live for more than 1400 years under controlled conditions (Jones et al. 2014). This amazing diversity suggests the multiple evolutionary origins of aging, making it difficult to develop an inclusive aging theory. Therefore, for better understanding of mechanisms underlying aging, if not comprehensive, attempts should be made to obtain knowledge from a wide variety of organisms.

Rotifers are the first model organism used in aging studies due to their several advantages such as short lifespan, ease of culture, and asexual reproduction that enables clonal culture (Enesco 1993). In addition to these advantages that are still valid today, molecular phylogenetic analyses revealed that rotifers belong to the superphylum Lophotrochozoa and phylogenetically distant from major invertebrate models, the nematode *Caenorhabditis elegans* and the fruit fly *Drosophila melanogaster* (Austad 2009). Although little molecular information has been available compared to these animals, recent advances in genome and transcriptome sequencing as well as development of gene knockdown techniques are going to make rotifers the traditional, but also up-to-date, model organism in aging studies. This chapter summarizes current knowledge of rotifer aging in association with some traditional aging theories and future prospects. Reevaluation of past works, as well as future advances in rotifer studies, would fill the taxonomic gaps and help us to understand aging mechanisms from a broad perspective. Several valuable reviews are available on this issue (King 1969; King and Miracle 1980; Enesco 1993; Snell 2014, Snell et al. 2015).

8.2 General Features of Aging and Lifespan in Rotifers

The first report on rotifer lifespan was Plate (1886), in which a monogonont rotifer *Hydatina senta* showed an average lifespan of 14 days (cited in Enesco 1993). Since then, lifespan of rotifers has been extensively investigated along with various interventions that affect lifespan (representative studies are shown in Table 8.1). Most rotifers lived for a few days to a month, and they seem to have a high ability to extend lifespan in response to environmental changes. Indeed, more than twofold extension of lifespan can be observed, which is not common in other invertebrate models without entering diapause or being genetically manipulated (Klass 1977; Le Bourg and Minois 2005; Heestand et al. 2013). Lifespan of vertebrates is usually less variable. The remarkable lifespan plasticity would be another advantage of rotifers as a model, especially when they are used for screening of compounds that alter lifespan. It is noted that a bdelloid rotifer *Philodina acuticornis* increased lifespan by 368% when treated with an indole derivative; the authors argued that this is the most pronounced life extension ever recorded in rotifers and similar invertebrate models like *Caenorhabditis elegans* (Poeggeler et al. 2010).

Rotifers generally show an age-dependent increase in mortality. The mortality rate doubling time has therefore been calculated as an indicator of aging in many

Table 8.1 Examples of rotifer lifespan studies

Species	Intervention	Lifespan increase (% control)	Significance	Reference
Highly cited papers[a]				
Asplanchna brightwelli	–	–	Cost of reproduction	Snell and King (1977)
Philodina sp.	Vitamin E	110%	Antioxidant-induced lifespan extension	Enesco and Verdone-Smith (1980)
Brachionus patulus	Temperature and CCR	Multiple groups; the maximum difference 175%	Combined effects of temperature and food level	Sarma and Rao (1991)
Brachionus plicatilis	IF	270%	Trade-off between lifespan and fecundity	Yoshinaga et al. (2000)
Asplanchna brightwelli	Temperature and IF	Up to 194% by low temperature up to 140% by IF	No effects of low temperature and IF on the length of post-reproductive period	Verdone-Smith and Enesco (1982)
Anuraeopsis fissa and others	–	Multiple groups	Importance of bacteria as food source in nature	Ooms-Wilms (1997)
Recent papers of interest				
Philodina acuticornis	IPAM (indole derivative; inhibitor of oxidative phosphorylation)	368%	The most pronounced life extension in rotifers	Poeggeler et al. (2010)
Brachionus plicatilis	IF	153% (mothers) 134% (offspring)	Transmission of acquired longevity	Kaneko et al. (2011)
Brachionus manjavacas	CCR and IF	Multiple groups	Large effects of reproductive mode and strain	Gribble and Mark Welch (2013), Gribble et al. (2014a)
Brachionus manjavacas	Inhibitor and RNAi knockdown on TOR/JNK pathways	Multiple groups	Identification of regulatory networks	Snell et al. (2014)
Brachionus plicatilis s.s.	Chemostat culture for 385 days	126%	Retardation of aging by laboratory culture (evolution of lifespan)	Smith and Snell (2014)

[a]Web of science database screening with "rotifer" and "lifespan" as keywords resulted in 85 hits on June 2014. Highly cited original papers of rotifers were selected manually

species: *Asplanchna brightwelli*, *A. girodi*, *Brachionus calyciflorus*, *B. plicatilis*, *Euchlanis dilatata*, *Keratella testudo*, and *Synchaeta pectinata* (Kirk 2001; Smith and Snell 2014).

In some species including *Philodina acuticornis* and *Macrotrachela quadricornifera*, mortality rate is markedly increased after the cessation of reproduction (Meadow and Barrows 1971; Ricci and Fascio 1995). This phenomenon used to be considered as evidence of programmed aging, where the genetically controlled biological clock determines aging rate and lifespan of animals (*programmed aging theory*). Because of the apparent lack of evolutionary benefit in programmed aging, non-programmed aging theories have dominated in the past few decades. Accordingly, aging of rotifers has been not well associated, at the molecular level, with reproduction in contrast to the close association with oxidative stress, which is considered to be a major cause of nonprogrammed aging (see *Oxidative stress and antioxidant defense* for details). However, the possible evolutionary benefits of limiting survival have been recently reevaluated in certain animals (Kirkwood and Melov 2011). Although it is not clear whether limiting survival is beneficial in rotifers, the strong coupling between reproduction and aging suggests the existence of negative lifespan regulation at least in some species. Meanwhile, in *C. elegans* several sets of genes involved in aging are transcriptionally regulated by "signals from reproductive system" (McCormick et al. 2012). Although molecules transducing this signal are not identified yet, they are suggested to be lipophilic hormones that diffuse through cellular membranes. Identification of the rotifer germline signals, if any, would provide new insight into *the programmed aging theory* and attract many researchers' attentions.

Decrease in swimming activity is another indicator of aging in rotifers as reported in *A. brightwelli* (Beauvais and Enesco 1985), *B. plicatilis* (Luciani et al. 1983), *B. manjavacas* (Snell et al. 2012), and *B. calyciflorus* (Yang et al. 2013a). Treatments that extend lifespan can ameliorate the decline of swimming activity in *B. manjavacas* (Snell et al. 2014). Rotifers also undergo other age-dependent changes such as body shrinkage, decrease in the size of nuclei, decrease in the rough endoplasmic reticulum, and increase in primary lysosomes (Enesco 1993), although a recent *B. manjavacas* study showed that the lysosome level is associated with reproduction rather than aging (Snell et al. 2014). These morphological changes implied the decline of protein synthesis and increase in protein degradation during aging, which would be reflected by low number of proteins in aged rotifers detected by SDS-PAGE (Carmona et al. 1989; Ricci et al. 1999). In addition, age-dependent calcium accumulation has been well documented in rotifers (King and Miracle 1980). The imbalance of calcium homeostasis might be involved in the above changes via some calcium-dependent pathways (Sawada and Carlson 1990).

Aging of monogonont males is completely different from that of females because males do not feed at all and die within a few days. The reported lifespan of males are 3.0 days for *B. plicatilis* (Snell and Childress 1987), 4.3 days for *B. manjavacas* (Snell 2014), and 3.1 days for *Epiphanes senta* (Schröder 2003). In this chapter we mostly describe about aging of amictic females unless stated.

8.3 Culture Temperature

Low culture temperature generally increases lifespan of rotifers as shown in *Philodina acuticornis* (Fanestil and Barrows 1965), *P. acuticornis* (Meadow and Barrows 1971), *Asplanchna brightwelli* (Verdone-Smith and Enesco 1982), *Synchaeta littoralis* (Bosque et al. 2001), *Brachionus patulus* (Sarma and Rao 1991), *B. calyciflorus* (Xi et al. 2005; Xiang et al. 2010; Kauler and Enesco 2011), *B. plicatilis* (King and Miracle 1980; Yoshinaga 2010), and *B. havanaensis* (Pavón-Meza et al. 2005). King and Miracle (1980) reported that a *B. plicatilis* clone had the same lifespan under 20 °C and 30 °C. This is unusual and interesting, but we found no subsequent publications on this clone.

The low temperature-induced longevity well supports *the rate of living theory*, a classic aging theory that postulates higher metabolism results in shorter lifespan. Although, to our knowledge, no one has directly measured metabolic flux of rotifers under different temperatures, reduced metabolism under low temperature has been suggested by many studies in terms of swimming speed (Larsen et al. 2008), intrinsic growth rate (Bosque et al. 2001), and changes in body composition under starved condition (Makridis and Olsen 1999). However, these findings do not indicate the causal relationship between reduced metabolism and longevity. Oxygen consumption was almost uniform from 20 to 28 °C in *B. plicatilis* hatched from naturally collected resting eggs (Epp and Lewis 1980), whereas a laboratory *B. plicatilis* strain had a significantly different lifespan under 20 °C and 25 °C (Yoshinaga 2010). Low temperature-induced longevity is thus not simply attributed to metabolic reduction although the interstrain difference might exist in the above example. Indeed, temperature-dependent changes of investment to lifespan and reproduction were suggested from a comparative study of morphologically distinct *B. calyciflorus* (two spined and unspined) (Xiang et al. 2010). Hormesis-like effect is likely to be involved too, because *B. calyciflorus* exposed to low temperature increased the mRNA levels of several genes related to stress response (Yang et al. 2014).

8.4 Calorie Restriction

Calorie restriction (CR; also called caloric restriction or dietary restriction) is a reduction of caloric intake without malnutrition. CR is known to retard the aging process in various animals including rotifers, and in most cases CR also extends lifespan. Two CR methods have been widely employed in rotifer aging studies: intermittent fasting (IF) where rotifers are fed at regular intervals and chronic calorie restriction (CCR) where rotifers are continuously fed with lower concentration of food. Also, in some CR protocols, rotifers were completely starved from the onset of food deprivation to when they died or were sampled for the following experiments (Kaneko et al. 2005; Yang et al. 2013a, b). It is confusing to regard these interventions as CR because the period of food deprivation was sometimes considerably long (up to 2–4 days in the papers cited above) and probably resulted in malnutrition. We would like to define such procedures as "starvation" here and the period that rotifers survived after the onset of food deprivation as "starvation time" or "survival time." Several authors used these terms (Kirk 1997; Yoshinaga et al. 2003; Gilbert 2004).

The effects of IF on lifespan have been well investigated in the genus *Brachionus*. Feeding 1 or 3 h a day increased lifespan of *B. plicatilis* up to 270% compared to controls fed with ad libitum (AL) (Yoshinaga et al. 2000). The CR and AL regimens correspond to the estimated feeding periods of rotifers in stationary phase and exponential growth phase, respectively, suggesting that lifespan of *B. plicatilis* varies according to the population growth phase. Alternate-day feeding also increased *B. plicatilis* lifespan (Kaneko et al. 2011). IF-induced longevity is also reported for *B. calyciflorus* (Ozdemir 2009; Yang et al. 2014), *B. manjavacas* (Gribble and Mark Welch 2013; Gribble et al. 2014a, b), *Philodina acuticornis* (Fanestil and Barrows 1965), and *Asplanchna brightwelli* (Verdone-Smith and Enesco 1982).

Significant lifespan extension was also observed in rotifers subjected to CCR: *P. acuticornis* (Meadow and Barrows 1971), *A. brightwelli* (Sawada and Enesco 1984a), *Elosa worallii* (Weithoff 2007), and *B. manjavacas* (Gribble and Mark Welch 2013). On the contrary, CCR did not increase lifespan of *Synchaeta littoralis* (Bosque et al. 2001), *B. calyciflorus* (Xi et al. 2001), *B. havanaensis* (Pavón-Meza et al. 2005), and *Cephalodella* sp. (Weithoff 2007). The difference in IF and CCR remains to be elucidated, but IF frequently extends lifespan than CCR in various animals (Kenyon 2010).

The CR-induced longevity is often associated with reproductive suppression in rotifers as demonstrated in *A. brightwelli* (Verdone-Smith and Enesco 1982), *B. plicatilis* (Yoshinaga et al. 2000; Kaneko et al. 2011), and *B. calyciflorus* (Ozdemir 2009; Yang et al. 2014) subjected to IF, as well as *P. acuticornis* (Meadow and Barrows 1971) under CCR. The trade-off between lifespan and fecundity has been interpreted as a reproductive strategy to postpone reproduction until environmental conditions improve for offspring. This interpretation is based on the assumption proposed by *the disposable soma theory*. Namely, there is a limited resource for reproduction and somatic maintenance, and thereby animals cannot reproduce actively when they invest a lot in somatic maintenance (i.e., when they have long lifespan under CR). For example, in a comparative study of two rotifer species *E. worallii* and *Cephalodella* sp., which occupy a similar niche in an extremely acidic (pH 2.7) mining lake, only the former species extended lifespan under 10% CCR (Weithoff 2007). Since *E. worallii*, but not *Cephalodella* sp., suppressed reproduction under 10% CCR, one would assume that the reduced fecundity enabled *E. worallii* to invest more in somatic maintenance. An interspecies comparison of starvation response revealed the similar trade-off between starvation time and reproduction among nine rotifer species (Kirk 1997). The negative effect of reproduction on lifespan was also demonstrated from variations of life history parameters in AL-fed *A. brightwelli* (Snell and King 1977).

However, several in-depth CR studies revealed the lack of trade-off between reproduction and lifespan. The IF and CCR differently influenced reproduction and lifespan in sexual and asexual *B. manjavacas*, and the trade-off was observed only when amictic females derived from amictic eggs were subjected to IF (Gribble and Mark Welch 2013). Probably CCR is prone to increase lifespan without the trade-off in rotifers because reproductive suppression was also accompanied with longevity induced by IF (Verdone-Smith and Enesco 1982) but not by CCR (Sawada and Enesco 1984a) in *A. brightwelli.*

The heterogeneous responses of rotifers to IF and CCR can be rationally attributed to the different modes of gene expression. So far, there has been only a small-scale screening for CR-induced genes (*B. plicatilis* under IF) (Oo et al. 2010), but a huge number of genes involved in IF and CCR responses will be identified in the next decade taking advantage of the next-generation sequencing technology. Molecular information, probably more than biological data, would facilitate distinguishing IF and CCR responses and comparing CR effects among species. It would also help us to answer the long-standing questions: "what is the fundamental response to CR?", "how CR responses have diverged in animals?", and "what is the substantial basis of trade-off and investment?". Nevertheless, it is noted that aging, and CR response with more probability, might have evolved independently in various taxa of rotifers (Gribble et al. 2014a). A considerable portion of CR response can be species-specific. Indeed, evolution of CR response was observed within a few hundred generations in a chemostat culture (Smith and Snell 2014).

8.5 Food Quality

Food quality affects lifespan of rotifers. *Brachionus manjavacas* fed with the feeding algae *Dunaliella tertiolecta* lived 20–25 % longer than those fed with *Tetraselmis suecica* (Snell 2014). The author further demonstrated that culture condition of *D. tertiolecta* also influenced *B. manjavacas* lifespan and argued that the biochemical composition of the algae, especially the protein/carbohydrate ratio, would be important for this life extension. It is a plausible and interesting explanation because protein restriction extends lifespan in *Drosophila melanogaster* (Sun et al. 2012) and mice (Goodrick 1978; Youngman et al. 1992) through mechanisms partially overlapping with those for CR.

8.6 Oxidative Stress and Antioxidant Defense

It is now widely accepted that aging is related to the oxidation of biomolecules by reactive oxygen species (ROS) produced as by-products of oxidative metabolism. This idea was originally proposed as *the free radical theory of aging* (Harman 1956). Since not only free radicals but also non-radical ROS such as hydrogen peroxide (H_2O_2) and singlet oxygen ($^1O_2$) cause the oxidative damage, this theory is now commonly called *the oxidative theory of aging*.

There is only fragmentary knowledge about the oxidative damage in rotifers, but they are basically consistent with this theory. The amount of lipid peroxide and malondialdehyde (lipid peroxidation products) increased with age in *Asplanchna brightwelli* (Sawada and Carlson 1985, 1990). The amount of malondialdehyde negatively correlated to lifespan when *A. brightwelli* was subjected to interventions that affect lifespan (e.g., low temperature and CR) (Sawada and Carlson 1990). Ultraviolet (UV) irradiation resulted in the accumulation of carbonylated proteins in *Adineta vaga* although the rate was much lower than that in *Caenorhabditis elegans* due to the high antioxidant protection of this rotifer species (Krisko et al. 2012).

The oxidative theory of aging predicts that (1) external oxidative stresses shorten lifespan and (2) antioxidants extend lifespan through the elimination of harmful ROS. Oxidative agents such as juglone and paraquat actually decreased lifespan of *Brachionus plicatilis* and *B. calyciflorus* (Kaneko et al. 2011; Tanaka et al. 2009). The latter prediction has also been verified at least in certain cases. Vitamin E or its water-soluble derivative trolox increased lifespan in *Philodina* sp. (Enesco and Verdone-Smith 1980) and *A. brightwelli* (Sawada and Enesco 1984b), but not in *B. calyciflorus* (Yang et al. 2014) and *B. manjavacas* (Snell et al. 2012). The ability of vitamin E to eliminate ROS was confirmed in *A. brightwelli*; it decreased lipid peroxidation caused by UV irradiation (Sawada et al. 1990). A nitrone compound α-phenyl-*N*-*tert*-butyl nitrone (PBN) and its amphiphilic derivative increased lifespan of a bdelloid rotifer *P. acuticornis* up to 250% (Poeggeler et al. 2005). Also, an indole derivative indolepropionamide (IPAM), which reduces electron leakage from the mitochondrial electron transport chain and thereby decreases ROS production, extended *P. acuticornis* lifespan up to 368% (Poeggeler et al. 2010).

What is important here is that only a few ROS scavengers have the ability to extend rotifer lifespan. Litton Jr. (1987) showed in four bdelloid rotifers that only limited analogs of vitamin E increased lifespan, suggesting that lifespan extension by vitamin E is not simply attributed to its antioxidant capacity. A recent large-scale screen failed to detect any significant lifespan extension in *B. manjavacas* exposed 20 antioxidants including trolox and IPAM (Snell et al. 2012). Among 60 two-way combinations of these antioxidants, only seven resulted in the significant life extension. In addition, three compounds screened from red algal extracts extended lifespan of *B. manjavacas* up to 114%, but their life-extending effects were not correlated to antioxidant capacities (Snare et al. 2013). Hence, oxidative damage and antioxidant defense system certainly affect lifespan of rotifers, but there must be something beyond them.

In addition to the low molecular weight antioxidants, organisms also use enzymatic defense system against oxidative stress. Manganese superoxide dismutase (Mn SOD) and catalase are two well-known antioxidant enzymes in rotifers. Mn SOD, an enzyme generally found in mitochondria, catalyzes the superoxide (O_2^-) to H_2O_2 and molecular oxygen (O_2), and catalase further degrades H_2O_2 to water (H_2O) and molecular oxygen. The mRNA and protein levels of these enzymes were increased by IF in *B. plicatilis* (Kaneko et al., 2011), providing the potential molecular basis of IF-induced longevity. The mRNA levels of both or either of these enzymes were increased by starvation (Kaneko et al. 2005; Yang et al. 2013) and by exposure to oxidative stress (Kailasam et al. 2011; Yang et al. 2013c) in *Brachionus* rotifers. A cytosolic isoform of SOD, copper-zinc SOD (Cu/Zn SOD), has been characterized in *B. plicatilis* (Denekamp et al. 2009) and *B. calyciflorus* (Yang et al. 2013c). This enzyme would also contribute to longevity because the mRNA levels were increased by starvation in *B. calyciflorus* (Yang et al. 2013c). However, the lack of causal knowledge limits the overall importance of these antioxidant enzymes in lifespan regulation. The above findings await further examination by transgenic and knockdown studies.

Four glutathione S-transferase (GST) genes have been cloned from *B. koreanus*, and their mRNA levels were increased in association with intracellular ROS induced by $CuCl_2$ treatments (Han et al. 2013). UV irradiation increased ROS levels and activities of GST, glutathione peroxidase (GPx), and glutathione reductase (GR) in *Brachionus* sp. (Kim et al. 2011). In *B. plicatilis*, GST genes were associated with antioxidant activity along with various heat shock protein and ferritin genes (Denekamp et al. 2009, 2011).

Interestingly, the mRNA levels of Mn SOD, catalase, and Cu/Zn SOD were increased by vitamin E in *B. calyciflorus* (Yang et al. 2013c). Assuming that these antioxidant enzymes are able to extend lifespan, we can consider that vitamin E is not a simple antioxidant but is a signaling molecule that activates longevity pathway(s). Moreover, it is also possible that increase in antioxidant enzymes itself triggers the activation of other longevity pathways; this hypothesis is motivated by the report that a *Drosophila melanogaster* strain overexpressing Mn SOD had low metabolic rate (CO_2 production) and a global gene expression pattern similar to that of a long-lived *C. elegans* strain with a mutation in the insulin-like signaling pathway (Curtis et al. 2007). This pathway is known to increase the expression of antioxidant enzymes and alter the whole-body energy balance toward longevity (see *Regulatory networks* for details). Therefore, lifespan seems to be regulated by complex feedback networks composed of ROS levels, antioxidant enzymes, and signaling pathways regulating whole-body energy metabolism. Further studies are required to portray their relationships in rotifers.

8.7 Metabolic Shift

Glycolysis is the sequential reaction that produces two pyruvate and two ATP molecules from one glucose molecule. Under normoxic conditions, majority of the resulting pyruvates (>95%) are converted to acetyl-CoA and further oxidized by the tricarboxylic acid (TCA) cycle in mitochondria, yielding NADH and $FADH_2$. Oxidative phosphorylation produces ATPs using these high-energy molecules. When less oxygen is available, on the other hand, pyruvate is converted to other metabolites such as lactate, ethanol, and succinate, and these reactions result in the production of less amount of ATP than oxidative phosphorylation. Although little information is as yet available, it is likely that rotifers use the above central metabolic pathways to utilize glucose as an energy source. The mRNA levels of glycogen phosphorylase, a rate-limiting enzyme of glucose-1-phosphase production from glycogen, were increased by CR (Oo et al. 2010). Production of lactate by lactate dehydrogenase (LDH) is the dominant reaction under hypoxia in *Brachionus plicatilis* (Esparcia et al. 1992).

Modulations of glucose metabolism affect lifespan of rotifers in several ways. Reduction of total glucose utilization, by inhibiting the first step of glycolysis with a hexokinase inhibitor 2-deoxyglucose, extends lifespan of *B. manjavacas* possibly due to the activation of CR pathways (Snell and Johnston 2014), suggesting that glucose availability is an index of nutritional condition in rotifers. Glycerol supplementation also increased lifespan of *B. manjavacas* by shifting metabolism from glycolysis to oxidative phosphorylation (i.e., utilization of alternative carbon source) (Snell and Johnston 2014). The general metabolic response to food deprivation is to switch the carbon source from glucose to other energy depots, primarily triacylglycerol (TAG), which is also oxidized by the TCA cycle. Since the catabolism of TAG yields fatty acids and glycerol, glycerol might deliver signals that trigger CR pathways. Glycerol also functions as a chemical chaperone (Perlmutter 2002) and improves the resistance to several stresses in *B. manjavacas* (Snell and Johnston 2014).

Apparently paradoxically, enhancement of glycolysis has also been associated with longevity in *B. plicatilis*. The mRNA levels of two glycolytic enzymes, glyceraldehyde 3-phosphate dehydrogenase (GAPDH), and enolase were increased by IF (Ozaki et al. 2010). This might be a simple reflection of limited glucose supply but can be a positive phenomenon to reduce oxidative stress under CR since overexpression of glycolytic enzymes is able to reduce oxidative damage in rodent cells (Kondoh et al. 2005). Although the mechanism by which enhancement of glycolysis reduces oxidative stress is currently unknown, it is probably associated with increase in the LDH flux (lactate production from glucose that does not use oxygen). Metabolomics studies would be useful to explore these issues.

8.8 Stress Response

Exposure to mild stress is known to extend lifespan of animals although a recent meta-analysis revealed that such a hormetic response takes place only under limited conditions (Lagisz et al. 2013). In rotifers, mild heat stress improved survival of *Brachionus plicatilis* upon the following severe heat treatment (Yoshinaga et al. 2008), but the effects of any mild stress on lifespan have not been examined yet.

Heat shock proteins (HSPs) are protein families rapidly synthesized in response to various kinds of stresses such as heat, heavy metals, and UV irradiation (Sanders 1993). They are commonly classified into HSP60, HSP70, HSP90, and low molecular weight (small) HSPs according to the molecular weight. The primary function of HSPs is to assist refolding or degradation of damaged proteins, and several HSPs are involved in lifespan extension induced by mild stress (Shama et al. 1998; Hartwig et al. 2009). Literatures about rotifer HSPs, again fragmentary, imply that they have similar functions with HSPs of other organisms. Rotifer HSPs were induced by various stresses at mRNA (Cochrane et al. 1994; Kim et al. 2011; Rhee et al. 2011; Jung and Lee 2012; Yang et al. 2014) and protein levels (Cochrane et al. 1991; Wheelock et al. 1999; Rios-Arana et al. 2005). It is noteworthy that causal relationship between HSP and stress resistance has been partly validated in rotifers; exogenous expression of *Brachionus* sp. HSP20 increased heat stress resistance of the host bacteria *Escherichia coli* (Rhee et al. 2011), and RNAi knockdown of HSPs in *B. manjavacas* reduced their thermotolerance (Smith et al. 2012). Expression of HSPs was increased by CR in *Brachionus* rotifers, suggesting that HSPs are involved in CR-induced longevity of rotifers (Wheelock et al. 2002; Yang et al. 2014). Meanwhile, the effects of vitamin E on expression of HSPs were not straightforward in *B. calyciflorus*; the mRNA levels of HSP60 and HSP70 were increased, whereas those of HSP40 and HSP90 were decreased (Yang et al. 2014). Taken together, we consider that the next challenge of rotifer HSP studies will be to find the regulatory mechanism of their expression and to reveal a causal relationship between HSPs and longevity.

8.9 Regulatory Networks

The insulin-like pathway is a conserved signaling system that regulates lifespan, stress resistance, energy metabolism, and reproduction in several phylogenetically distant animals. The loss-of-function mutations in components of this pathway led to the extension of lifespan in *Caenorhabditis elegans*, *Drosophila melanogaster*, and mice (Kenyon 2010). In human, at least 14 different mutations of this pathway have been associated with longevity (Kenyon 2010). Moreover, the transcription of Mn SOD, catalase, and low molecular weight HSP genes is regulated by this pathway in *C. elegans* (Murphy et al. 2003). Collectively, it is highly plausible that the insulin-like signaling pathway has some impacts on rotifer lifespan.

Although only few data are available, one could assume that CR decreases the activity of this pathway and thereby extends lifespan of *B. plicatilis* from the following observations. First, aqueous extract of *B. plicatilis* activated mammalian insulin-like signaling pathway, and extracts from fed rotifers resulted in greater activation than those from starved rotifers (Ozaki et al. 2013). Second, the LY294002 inhibitor of phosphatidylinositol 3-kinase, a component of the insulin-like signaling pathway, increased lifespan of *B. plicatilis* up to 130% (Yoshinaga et al. 2005). Third, CR increased the mRNA levels of Mn SOD (Kaneko et al. 2011), the transcription of which is negatively regulated by the insulin-like pathway in *C. elegans* and mammals (Kops et al. 2002; Murphy et al. 2003). However, another PI3K inhibitor AS-605240 as well as RNAi knockdown of PI3K failed to extend lifespan of *B. manjavacas* although there is room for discussion about specificity of these inhibitors and knockdown efficiency (Snell et al. 2014). Further research is necessary to clarify the role of this pathway in rotifer lifespan.

The target of rapamycin (TOR) is a protein kinase that integrates a variety of signaling networks involved in nutrient conditions, growth, energy metabolism, and stress response. The TOR pathway is known to interact with a well-conserved stress response pathway, the Jun-N-terminal kinase (JNK) pathway. Inhibitors of TOR and JNK increased *B. manjavacas* lifespan and starvation time in an additive manner, but did not affect the resistance to heat and oxidative stresses (Snell et al. 2014). RNAi knockdown of TOR, JNK, and their downstream kinase Akt also extended lifespan, clearly demonstrating that these pathways mediate lifespan of *B. manjavacas*. It is noted that Akt is a common component of TOR and insulin-like signaling pathways in many animals. Given that TOR and JNK inhibition did not increase stress resistance, interaction of several signaling pathways should be taken into account.

8.10 Diapause and Maternal Effect

Maternal conditions substantially influence life history of offspring in rotifers. The most prominent maternal effect is the production of resting eggs by mictic females. Resting eggs can survive for decades possibly without affecting the post-hatching lifespan. For instance, *B. manjavacas* hatched from resting eggs stored for 23 or 26 years had the same lifespan with those hatched resting eggs kept for 4 years (Snell 2014). This is similar to the case of *Caenorhabditis elegans*, in which the duration of developmentally arrested "dauer" stage did not affect the post-dauer lifespan (Klass and Hirsh 1976).

Also famous is the maternal age effect, where offspring from older mothers live shorter. It was first revealed in human by a survey of children's life expectancy in the early twentieth century and then experimentally indicated in the rotifer *Proales sordida* by Jenning and Lynch in 1928 (cited in King and Miracle 1980). Using isogenic lines of the rotifer *Philodina citrina*, Lansing (1947) further demonstrated that this tendency is transmitted to successive generations in a cumulative manner (Lansing 1947). He also showed that the introduction of young mother in the isogenic lines of late-born offspring canceled the accumulation of deleterious maternal age effects. This transmissible, cumulative, and reversible maternal age effect is called "Lansing effect" based on this widely quoted work.

Kaneko et al. (2011) demonstrated for the first time that CR-induced longevity can be transmitted to the next generation in *Brachionus plicatilis* (Kaneko et al. 2011); offspring from IF mothers had longer lifespan associated with higher oxidative stress resistance and higher catalase mRNA levels at birth compared to offspring from AL mothers. Yoshinaga et al. (2001) demonstrated that maternal IF increases starvation tolerance of offspring. These would be interpreted as a variation of the Lansing effect because maternal IF rescued the deleterious effects of advanced maternal age on offspring (Gribble et al. 2014b).

We envision two possible mechanisms for this maternal age effects: (1) age-dependent modification of DNA or histone proteins that affect chromatin structure and gene expression and (2) accumulation of the "aging factor(s)" in germline cells during mother's aging process. The DNA and histone modification is being actively investigated as a central mechanism of such transgenerational effects on longevity (Greer et al. 2011). However, modification (mainly methylation) of DNA and histone seems to be an "active and programmed" process that requires certain enzymes, in contrast to the "passive and non-programmed" accumulation of the aging factor(s). Currently we cannot find any adaptive significance in actively limiting offspring quality by the maternal age-dependent modification of DNA or histone.[1] Accumulation and transmission of aging factors, on the other hand, have been proved in a budding yeast *Saccharomyces cerevisiae* (Sinclair and Guarente 1997), but it is unclear whether such aging factors, if any, accumulate in germline cell of multicellular animals like rotifers. Clearly, further studies are needed to understand the molecular basis of the maternal age effects. Meanwhile, young and old mothers of a bdelloid rotifer *Macrotrachela quadricornifera* had distinct SDS-PAGE protein band patterns, and the difference was reproduced in offspring from young and old mothers (Ricci et al. 1999). The difference in protein composition can explain the phenotypic variation, and both of the abovementioned mechanisms are capable to change the protein composition in offspring.

8.11 Conclusion

Biology is frequently referred to as "the science of exceptions." Rotifers have indeed provided several exceptions in traditional aging theories such as odd aging of males, lack of trade-off between reproduction and somatic maintenance, ambiguous effects of antioxidants on lifespan, and transmission of acquired longevity to the next generation. These findings, however, probably do not mean that rotifers are extremely unusual animals; rather, they merely suggest the diversity of mechanisms underlying the aging process, and only recently researchers started to acknowledge the messages from rotifers. We suppose that the aging study, which has been developed based on information from a limited number of animals, is now in an age of expansion with more and more exceptions potentially found from diverse animals in the near future. Beyond the diversity, researchers might understand the fundamental and species-specific mechanism of aging. We hope that the knowledge from rotifers will be a milestone in the new epoch.

[1] During the revision of this chapter, van den Heuvel et al. (2016) published a simulation study showing the maternal age effect could be adaptive. PLoS ONE 11(1): e0145544.

References

Austad SN (2009) Is there a role for new invertebrate models for aging research? J Gerontol Ser A Biol Med Sci 64:192–194
Beauvais JE, Enesco HE (1985) Life span and age-related changes in activity level of the rotifer *Asplanchna brightwelli*: influence of curare. Exp Gerontol 20:359–366
Bosque T, Hernandez R, Pérez R, Todoli R, Oltra R (2001) Effects of salinity, temperature and food level on the demographic characteristics of the seawater rotifer *Synchaeta littoralis* Rousselet. J Exp Mar Biol Ecol 258:55–64
Carmona MJ, Serra M, Miracle MR (1989) Protein patterns in rotifers: the timing of aging. Hydrobiologia 186:325–330
Cochrane BJ, Irby RB, Snell TW (1991) Effects of copper and tributyltin on stress protein abundance in the rotifer *Brachionus plicatilis*. Comp Biochem Physiol 98C:385–390
Cochrane BJ, Mattley YD, Snell TW (1994) Polymerase chain-reaction as a tool for developing stress protein probes. Environ Toxicol Chem 13:1221–1229
Curtis C, Landis GN, Folk D, Wehr NB, Hoe N et al (2007) Transcriptional profiling of MnSOD-mediated lifespan extension in *Drosophila* reveals a species-general network of aging and metabolic genes. Genome Biol 8:R262
Denekamp NY, Thorne MAS, Clark MS, Kube M, Reinhardt R et al (2009) Discovering genes associated with dormancy in the monogonont rotifer *Brachionus plicatilis*. BMC Genomics 10:108
Denekamp NY, Reinhardt R, Albrecht MW, Drungowski M, Kube M, Lubzens E (2011) The expression pattern of dormancy-associated genes in multiple life-history stages in the rotifer *Brachionus plicatilis*. Hydrobiologia 662:51–63
Enesco HE (1993) Rotifers in aging research: use of rotifers to test various theories of aging. Hydrobiologia 255:59–70
Enesco HE, Verdone-Smith C (1980) α-tocopherol increases lifespan in the rotifer *Philodina*. Exp Gerontol 15:335–338
Epp RW, Lewis WM Jr (1980) Metabolic uniformity over the environmental temperature range in *Brachionus plicatilis* (Rotifera). Springer, Rotatoria, pp 145–147
Esparcia A, Serra M, Miracle MR (1992) Relationships between oxygen concentration and patterns of energy-metabolism in the rotifer *Brachionus-plicatilis*. Comp Biochem Physiol B-Biochem Mol Biol 103:357–362
Fanestil DD, Barrows CH (1965) Aging in the rotifer. J Gerontol 20:462–469
Gilbert JJ (2004) Females from resting eggs and parthenogenetic eggs in the rotifer *Brachionus calyciflorus*: lipid droplets, starvation resistance and reproduction. Freshwat Biol 49:1505–1515
Goodrick CL (1978) Body weight increment and length of life: the effect of genetic constitution and dietary protein. J Gerontol 33:184–190
Greer EL, Maures TJ, Ucar D, Hauswirth AG, Mancini E et al (2011) Transgenerational epigenetic inheritance of longevity in *Caenorhabditis elegans*. Nature 479:365–371
Gribble KE, Mark Welch DB (2013) Life-span extension by sscaloric restriction is determined by type and level of food reduction and by reproductive mode in *Brachionus manjavacas* (Rotifera). J Gerontol Ser A Biol Med Sci 68:349–358
Gribble KE, Kaido O, Jarvis G, Mark Welch DB (2014a) Patterns of intraspecific variability in the response to caloric restriction. Exp Gerontol 51:28–37
Gribble KE, Jarvis G, Bock M, Mark Welch DB (2014b) Maternal caloric restriction partially rescues the deleterious effects of advanced maternal age on offspring. Aging Cell 13:623–630
Han J, Won E-J, Hwang D-S, Rhee J-S, Kim I-C et al (2013) Effect of copper exposure on GST activity and on the expression of four *GSTs* under oxidative stress condition in the monogonont rotifer, *Brachionus koreanus*. Comp Biochem Physiol C: Toxicol Pharmacol 158:91–100
Harman D (1956) Aging: a theory based on free radical and radiation chemistry. J Gerontol 11:298–300

Hartwig K, Heidler T, Moch J, Daniel H, Wenzel U (2009) Feeding a ROS-generator to Caenorhabditis elegans leads to increased expression of small heat shock protein HSP-16.2 and hormesis. Genes Nutr 4:59–67

Heestand BN, Shen Y, Liu W, Magner DB, Storm N et al (2013) Dietary restriction induced longevity is mediated by nuclear receptor NHR-62 in *Caenorhabditis elegans*. PLoS Genet 9:e1003651

Jones OR, Scheuerlein A, Salguero-Gómez R, Camarda CG, Schaible R et al (2014) Diversity of ageing across the tree of life. Nature 505:169–173

Jung M-Y, Lee Y-M (2012) Expression profiles of heat shock protein gene families in the monogonont rotifer *Brachionus koreanus*—exposed to copper and cadmium. Toxicol Environ Heal Sci 4:235–242

Kailasam M, Kaneko G, Oo AKS, Ozaki Y, Thirunavukkarasu AR et al (2011) Effects of calorie restriction on the expression of manganese superoxide dismutase and catalase under oxidative stress conditions in the rotifer *Brachionus plicatilis*. Fish Sci 77:403–409

Kaneko G, Yoshinaga T, Yanagawa Y, Kinoshita S, Tsukamoto K et al (2005) Molecular characterization of Mn-superoxide dismutase and gene expression studies in dietary restricted *Brachionus plicatilis* rotifers. Hydrobiologia 546:117–123

Kaneko G, Yoshinaga T, Yanagawa Y, Ozaki Y, Tsukamoto K et al (2011) Calorie restriction-induced maternal longevity is transmitted to their daughters in a rotifer. Funct Ecol 25:209–216

Kauler P, Enesco HE (2011) The effect of temperature on life history parameters and cost of reproduction in the rotifer *Brachionus calyciflorus*. J Freshwat Ecol 26:399–408

Kenyon CJ (2010) The genetics of ageing. Nature 464:504–512

Kim R-O, Rhee J-S, Won E-J, Lee K-W, Kang C-M et al (2011) Ultraviolet B retards growth, induces oxidative stress, and modulates DNA repair-related gene and heat shock protein gene expression in the monogonont rotifer, *Brachionus* sp. Aquat Toxicol 101:529–539

King CE (1969) Experimental studies on ageing in rotifers. Exp Gerontol 4:63–79

King CE, Miracle M (1980) A perspective on aging in rotifers. Hydrobiologia 73:13–19

Kirk KL (1997) Life-history responses to variable environments: starvation and reproduction in planktonic rotifers. Ecology 78:434–441

Kirk KL (2001) Dietary restriction and aging: comparative tests of evolutionary hypotheses. J Gerontol Ser A Biol Med Sci 56:B123–B129

Kirkwood TB, Melov S (2011) On the programmed/non-programmed nature of ageing within the life history. Curr Biol 21:R701–R707

Klass MR (1977) Aging in nematode *Caenorhabditis-elegans* – major biological and environmental-factors influencing life-span. Mech Ageing Dev 6:413–429

Klass MR, Hirsh D (1976) Non-ageing developmental variant of *Caenorhabditis elegans*. Nature 260:523–525

Kondoh H, Lleonart ME, Gil J, Wang J, Degan P et al (2005) Glycolytic enzymes can modulate cellular life span. Cancer Res 65:177–185

Kops GJ, Dansen TB, Polderman PE, Saarloos I, Wirtz KW et al (2002) Forkhead transcription factor FOXO3a protects quiescent cells from oxidative stress. Nature 419:316–321

Krisko A, Leroy M, Radman M, Meselson M (2012) Extreme anti-oxidant protection against ionizing radiation in bdelloid rotifers. Proc Natl Acad Sci U S A 109:2354–2357

Lagisz M, Hector KL, Nakagawa S (2013) Life extension after heat shock exposure: assessing meta-analytic evidence for hormesis. Ageing Res Rev 12:653–660

Lansing AI (1947) A transmissiblet, cumulative, and reversible factor in aging. J Gerontol 2:228–239

Larsen PS, Madsen CV, Riisgård HU (2008) Effect of temperature and viscosity on swimming velocity of the copepod *Acartia tonsa*, brine shrimp *Artemia salina* and rotifer *Brachionus plicatilis*. Aquat Biol 4:47–54

Le Bourg É, Minois N (2005) Does dietary restriction really increase longevity in *Drosophila melanogaster*? Ageing Res Rev 4:409–421

Litton JR Jr (1987) Specificity of the α-tocopherol (vitamin E) effect on lifespan and fecundity of bdelloid rotifers. Hydrobiologia 147:135–139
Luciani A, Chassé J-L, Clément P (1983) Aging in *Brachionus plicatilis*: the evolution of swimming as a function of age at two different calcium concentrations. Hydrobiologia 104:141–146
Makridis P, Olsen Y (1999) Protein depletion of the rotifer *Brachionus plicatilis* during starvation. Aquaculture 174:343–353
McCormick M, Chen K, Ramaswamy P, Kenyon C (2012) New genes that extend *Caenorhabditis elegans*' lifespan in response to reproductive signals. Aging Cell 11:192–202
Meadow ND, Barrows CH (1971) Studies on aging in a bdelloid rotifer. II. The effects of various environmental conditions and maternal age on longevity and fecundity. J Gerontol 26:302–309
Murphy CT, McCarroll SA, Bargmann CI, Fraser A, Kamath RS et al (2003) Genes that act downstream of DAF-16 to influence the lifespan of *Caenorhabditis elegans*. Nature 424:277–284
Oo AKS, Kaneko G, Hirayama M, Kinoshita S, Watabe S (2010) Identification of genes differentially expressed by calorie restriction in the rotifer (*Brachionus plicatilis*). J Comp Physiol 180B:105–116
Ooms-Wilms AL (1997) Are bacteria an important food source for rotifers in eutrophic lakes?. J Plank Res 19(8):1125–1141
Ozaki Y, Kaneko G, Yanagawa Y, Watabe S (2010) Calorie restriction in the rotifer *Brachionus plicatilis* enhances hypoxia tolerance in association with the increased mRNA levels of glycolytic enzymes. Hydrobiologia 649:267–277
Ozaki Y, Kaneko G, Hakuno F, Takahashi S-I, Watabe S (2013) Insulin/insulin-like growth factor-like activity in the aqueous extracts of the rotifer *Brachionus plicatilis*. Fish Sci 79:47–53
Ozdemir N (2009) The effect of caloric restriction on the life span and reproduction of fresh water rotifer (*Brachionus calyciflorus*). J Anim Vet Adv 8:669–673
Pavón-Meza EL, Sarma S, Nandini S (2005) Combined effects of algal (*Chlorella vulgaris*) food level and temperature on the demography of *Brachionus havanaensis* (Rotifera): a life table study. Hydrobiologia 546:353–360
Perlmutter DH (2002) Chemical chaperones: a pharmacological strategy for disorders of protein folding and trafficking. Pediatr Res 52:832–836
Poeggeler B, Durand G, Polidori A, Pappolla MA, Vega-Naredo I et al (2005) Mitochondrial medicine: neuroprotection and life extension by the new amphiphilic nitrone LPBNAH1 acting as a highly potent antioxidant agent. J Neurochem 95:962–973
Poeggeler B, Sambamurti K, Siedlak SL, Perry G, Smith MA et al (2010) A novel endogenous indole protects rodent mitochondria and extends rotifer lifespan. PLoS One 5:e10206
Rhee J-S, Kim R-O, Choi H-G, Lee J, Lee Y-M et al (2011) Molecular and biochemical modulation of heat shock protein 20 (*Hsp20*) gene by temperature stress and hydrogen peroxide (H2O2) in the monogonont rotifer, *Brachionus* sp. Comp Biochem Physiol C: Toxicol Pharmacol 154:19–27
Ricci C, Fascio U (1995) Life-history consequences of resource allocation of two bdelloid rotifer species. Hydrobiologia 299:231–239
Ricci C, Santo N, Radaelli E, Bolzern AM (1999) Epigenetic inheritance systems in bdelloid rotifers. I. Maternal-age-related biochemical effects. Ital J Zool 66:333–339
Rios-Arana JV, Gardea-Torresdey JL, Webb R, Walsh EJ (2005) Heat shock protein 60 (HSP60) response of *Plationus patulus* (Rotifera: Monogononta) to combined exposures of arsenic and heavy metals, pp 577–585
Sanders BM (1993) Stress proteins in aquatic organisms – an environmental perspective. Crit Rev Toxicol 23:49–75
Sarma S, Rao TR (1991) The combined effects of food and temperature on the life history parameters of *Brachionus patulus* MULLER (Rotifera). Int Rev Ges Hydrobiol 76:225–239
Sawada M, Carlson JC (1985) Association of lipid peroxidation during luteal regression in the rat and natural aging in the rotifer. Exp Gerontol 20:179–186
Sawada M, Carlson JC (1990) Biochemical changes associated with the mechanism controlling superoxide radical formation in the aging rotifer. J Cell Biochem 44:153–165

Sawada M, Enesco HE (1984a) A study of dietary restriction and lifespan in the rotifer *Asplanchna brightwelli* monitored by chronic neutral red exposure. Exp Gerontol 19:329–334
Sawada M, Enesco HE (1984b) Vitamin E extends lifespan in the short-lived rotifer *Asplanchna brightwelli*. Exp Gerontol 19:179–183
Sawada M, Carlson JC, Enesco HE (1990) The effects of UV radiation and antioxidants on life span and lipid peroxidation in the rotifer *Asplanchna brightwelli*. Arch Gerontol Geriatr 10:27–36
Schröder T (2003) Precopulatory mate guarding and mating behaviour in the rotifer *Epiphanes senta* (Monogononta: Rotifera). Proc R Soc Lond Ser B Biol Sci 270:1965–1970
Shama S, Lai C-Y, Antoniazzi JM, Jiang JC, Jazwinski SM (1998) Heat stress-induced life span extension in yeast. Exp Cell Res 245:379–388
Sinclair DA, Guarente L (1997) Extrachromosomal rDNA circles—a cause of aging in yeast. Cell 91:1033–1042
Smith HA, Snell TW (2014) Differential evolution of lifespan and fecundity between asexual and sexual females in a benign environment. Int Rev Hydrobiol 99:117–124
Smith HA, Burns AR, Shearer TL, Snell TW (2012) Three heat shock proteins are essential for rotifer thermotolerance. J Exp Mar Biol Ecol 413:1–6
Snare DJ, Fields AM, Snell TW, Kubanek J (2013) Lifespan extension of rotifers by treatment with red algal extracts. Exp Gerontol 48:1420–1427
Snell TW (2014) Rotifers as models for the biology of aging. Int Rev Hydrobiol 99:84–95
Snell TW, Childress M (1987) Aging and loss of fertility in male and female *Brachionus plicatilis* (Rotifera). Int J Inver Rep Dev 12:103–110
Snell TW, Johnston RK (2014) Glycerol extends lifespan of *Brachionus manjavacas* (Rotifera) and protects against stressors. Exp Gerontol 57:47–56
Snell TW, King CE (1977) Lifespan and fecundity patterns in rotifers: the cost of reproduction. Evolution:882–890
Snell TW, Fields AM, Johnston RK (2012) Antioxidants can extend lifespan of *Brachionus manjavacas* (Rotifera), but only in a few combinations. Biogerontology 13:261–275
Snell TW, Johnston RK, Rabeneck B, Zipperer C, Teat S (2014) Joint inhibition of TOR and JNK pathways interacts to extend the lifespan of *Brachionus manjavacas* (Rotifera). Exp Gerontol 52:55–69
Snell TW, Johnston RK, Gribble KE, Mark Welch DB (2015) Rotifers as experimental tools for investigating aging. Invertebr Reprod Dev 59(S1):5–10
Sun X, Komatsu T, Lim J, Laslo M, Yolitz J et al (2012) Nutrient-dependent requirement for SOD1 in lifespan extension by protein restriction in *Drosophila melanogaster*. Aging Cell 11:783–793
Tanaka C, Hashimoto Y, Nakao S, Yoshinaga T (2009) Effect of juglone on the survival time of two *Brachionus species* (Rotifera): species-specific tolerance against oxidative stress. Fish Sci 75:191–194
Verdone-Smith C, Enesco HE (1982) The effect of temperature and of dietary restriction on lifespan and reproduction in the rotifer *Asplanchna brightwelli*. Exp Gerontol 17:255–262
Weithoff G (2007) Dietary restriction in two rotifer species: the effect of the length of food deprivation on life span and reproduction. Oecologia 153:303–308
Wheelock CE, Wolfe MF, Olsen H, Tjeerdema RS, Sowby ML (1999) Hsp60-induced tolerance in the rotifer Brachionus plicatilis exposed to multiple environmental contaminants. Arch Environ Contam Toxicol 36:281–287
Wheelock CE, Baumgartner TA, Newman JW, Wolfe MF, Tjeerdema RS (2002) Effect of nutritional state on Hsp60 levels in the rotifer *Brachionus plicatilis* following toxicant exposure. Aquat Toxicol 61:89–93
Xi Y-L, Huang X-F, Jin H-J, Liu J-K (2001) The effect of food concentration on the life history of three types of *Brachionus calyciflorus* females. Int Rev Hydrobiol 86:211–217
Xi Y-L, Ge Y-L, Chen F, Wen X-L, Dong L-L (2005) Life history characteristics of three strains of *Brachionus calyciflorus* (Rotifera) at different temperatures. J Freshwat Ecol 20:707–713

Xiang X-L, Xi Y-L, Zhang J-Y, Ma Q, Wen X-L (2010) Effects of temperature on survival, reproduction, and Morphotype in offspring of two *Brachionus calyciflorus* (Rotifera) Morphotypes. J Freshwat Ecol 25:9–18

Yang J, Dong S, Jiang Q, Kuang T, Huang W et al (2013a) Changes in expression of manganese superoxide dismutase, copper and zinc superoxide dismutase and catalase in *Brachionus calyciflorus* during the aging process. PLoS One 8:e57186

Yang J, Dong S, Zhu H, Jiang Q, Yang J (2013b) Molecular and expression analysis of manganese superoxide dismutase (Mn–SOD) gene under temperature and starvation stress in rotifer *Brachionus calyciflorus*. Mol Biol Rep 40:2927–2937

Yang J, Dong S, Jiang Q, Si Q, Liu X et al (2013c) Characterization and expression of cytoplasmic copper/zinc superoxide dismutase (CuZn SOD) gene under temperature and hydrogen peroxide (H2O2) in rotifer *Brachionus calyciflorus*. Gene 518:388–396

Yang J, Mu Y, Dong S, Jiang Q, Yang J (2014) Changes in the expression of four heat shock proteins during the aging process in *Brachionus calyciflorus* (rotifera). Cell Stress, Chaperones 19:33–52

Yoshinaga T (2010) Individual life history and population dynamics in the rotifer *Brachionus plicatilis*. Nippon Suisan Gakkaishi 76:617–620. (in Japanese)

Yoshinaga T, Hagiwara A, Tsukamoto K (2000) Effect of periodical starvation on the life history of *Brachionus plicatilis* O. F. Müller (Rotifera): a possible strategy for population stability. J Exp Mar Biol Ecol 253:253–260

Yoshinaga T, Hagiwara A, Tsukamoto K (2001) Effect of periodical starvation on the survival of offspring in the rotifer *Brachionus plicatilis*. Fisheries Science 67(2):373–374

Yoshinaga T, Hagiwara A, Tsukamoto K (2003) Life history response and age-specific tolerance to starvation in *Brachionus plicatilis* O. F. Müller (Rotifera). J Exp Mar Biol Ecol 287:261–271

Yoshinaga T, Kaneko G, Kinoshita S, Furukawa S, Tsukamoto K, Watabe S (2005) Insulin-like signaling pathway involved in regulating longevity of the rotifer. Hydrobiologia 546:347–352

Yoshinaga T, Tanaka C, Yasuda H (2008) Monitoring of the physiological status as the heat stress tolerance in the rotifer: a novel test for health diagnosis and ecological monitoring. Coast Mar Sci 32:30–32

Youngman LD, Park J, Ames BN (1992) Protein oxidation associated with aging is reduced by dietary restriction of protein or calories. Proc Natl Acad Sci U S A 89:9112–9116

Chapter 9
Using Rotifers to Diagnosis the Ecological Impacts of Toxicants

Terry W. Snell and Helen S. Marcial

Abstract Rotifers continue to be useful in assessing toxicity in a variety of applications. Substantive advances on several ecotoxicological topics have been made since the last reviews of rotifers in ecotoxicology in 2011 and 2013. We review advances in understanding the impact of endocrine disruptors on rotifers, how bioconcentration of lead affects aquatic food webs, nanoparticle toxicity to rotifers, the toxicity of fine aerosols in the atmosphere, the toxicity of metabolites from human pharmaceuticals, and how toxicants modify the immune response of rotifers and their response to pathogens. Also discussed is the use of other rotifer species besides *Brachionus* in toxicity studies, as well as the presence of hormesis in rotifer dose-responses. The use of rotifers in drug discovery efforts screening for bioactive natural compounds is described, as is employing rotifers to quantify the toxicity of harmful algae blooms. Finally, we describe how rotifer toxicity tests were employed to assess the toxicity of crude oil spills in the Gulf of Mexico and the consequences of treating spills with oil dispersants. All of these advances serve as examples that demonstrate how the use of rotifers in ecotoxicological studies continues to be an active and interesting area of research.

9.1 Introduction

The use of rotifers to diagnose the ecological impacts of natural and anthropogenic compounds continues to expand. The last reviews of rotifers and ecotoxicology were published by Dahms et al. (2011) and Rico-Martínez et al. (2013a). They emphasized gene expression studies, the toxicity of pharmaceuticals, endocrine disruption, and general ecotoxicological studies. In this review, we focus on papers published since 2011 and on topical areas where we believe that interesting experimental advances

---

T.W. Snell (✉)
School of Biological Sciences, Georgia Institute of Technology,
Atlanta, GA 30332-0230, USA
e-mail: terry.snell@biology.gatech.edu

H.S. Marcial
SEAFDEC/AQD, Tigbauan, Iloilo 5021, Philippines
e-mail: hsmarcial@seafdec.org.ph

A. Hagiwara, T. Yoshinaga (eds.), *Rotifers*, Fisheries Science Series,
DOI 10.1007/978-981-10-5635-2_9

are being made. These areas include endocrine disruption, bioconcentration, nanoparticles, toxicity of fine aerosols, pharmaceuticals, rotifer immune response to toxicants, new rotifer species in ecotoxicology studies, hormesis, screening natural compounds, harmful algal blooms, and assessing the toxicity of crude oil spills.

9.2 Endocrine Disruption

Endocrine-disrupting chemicals (EDCs) are hormonally active substances that act as agonists or antagonists to hormone receptors or, in other indirect ways, to perturb endocrine systems (Vos et al. 2000). EDCs include organic chemicals such as dioxins and dioxin-like compounds, pharmaceuticals, some metals, polychlorinated biphenyls, organochlorine pesticides, synthetic steroids, plasticizers, and surfactants. Several studies showed abnormalities in hormonally regulated functions including reproduction and development both of vertebrates and invertebrates when exposed to EDCs.

Although the endocrine systems regulating rotifer growth and reproduction remain obscure, various phases of reproduction, including asexual and sexual egg reproduction, sexual female reproduction, and sperm and egg maturation, are thought to be mediated through endocrine mechanisms (Snell 2011). Recently, progesterone and a vitellogenin-like protein have been detected in the rotifer *Brachionus manjavacas* (Stout et al. 2010; Jones et al. 2013). Furthermore, Alvarado-Flores et al. (2009) found immunoreactivity in females, males, and eggs of *Brachionus calyciflorus* when exposed to mammalian-luteinizing hormone, follicle-stimulating hormone, thyroid-stimulating hormone, and prolactin (Table 9.1.). Thus, the anomalies observed in rotifers due to the presence of EDCs in their culture medium is indeed not primarily due to mere toxicity, but rather mediated by their hormonal system. Dahms et al. (2011) reviewed studies that showed that rotifers are sensitive to a variety of EDCs. Recent studies also support their claim. For example, García-García et al. (2014) found that exposure to low concentrations of combined ethinylestradiol and levonorgestrel significantly enhanced the intrinsic rate of population increase and fecundity of *Anuraeopsis fissa* and *B. calyciflorus*. These two synthetic hormones are the main components of contraceptives in Mexico (García-García et al. 2014) and are known to cause endocrine disruption in aquatic animals (Duffy et al. 2014). Also, phthalate esters such as di-*n*-butyl phthalate (DBP) and butyl benzyl phthalate (BBP) which are commonly used in PVCs have been shown to increase the net reproductive rate and prolong the generation time and life expectancy upon hatching of *B. calyciflorus* when added at a relatively high concentration (500 μg/l), while at lower concentrations (50 μg/l), the intrinsic rate of population increase was significantly decreased (Zhao et al. 2009). Organochlorine pesticides including aldrin, dieldrin, β-hexachlorocyclohexane (β-HCH) and chlordecone also caused reproductive anomalies in *B. calyciflorus*. Huang et al. (2012) found that addition of low concentrations (0.01–10 μg/L) of dieldrin and 17β-estradiol can shorten the duration of embryonic development and increase the reproductive

Table 9.1 Summary of species, toxicants, and endpoints in rotifer studies

Species	Toxicants	Endpoints	Reference
Brachionus calyciflorus	Luteinizing hormone, follicle-stimulating hormone, thyroid-stimulating hormone, prolactin	Immunoreactivity	Alvarado-Flores et al. (2009)
B. calyciflorus, *Anuraeopsis fissa*	Ethinylestradiol and levonorgestrel	Population growth rate, fecundity	García-García et al. (2014)
B. calyciflorus	di-*n*-butyl phthalate, butyl benzyl phthalate	Population growth rate, generation time, life expectancy	Zhao et al. (2009)
B. calyciflorus	Aldrin, dieldrin, β-hexachlorocyclohexane (β-HCH), chlordecone, 17β-estradiol	Development rate, reproductive and post-reproductive period, net reproduction, generation time, life expectancy	Huang et al. (2012)
Asplanchna brightwellii	Lead	Bioconcentration	Rubio-Franchini and Rico-Martinez (2011) and Alvarado-Flores et al. (2012)
B. calyciflorus	Lead	Bioconcentration	Alvarado-Flores et al. (2012)
Euchlanis dilatata	Lead	Mortality	Arias-Almeida and Rico-Martínez (2011a, b)
Philodina rapida	Lead	Population growth rate	Esbaugh et al. (2012)
B. calyciflorus	Lead	Population growth rate	Grosell et al. (2006)
B. manjavacas	Nanoparticles	Population growth rate	Snell and Hicks (2009)
B. plicatilis	TiO_2 nanoparticles	Immobility	Clément et al. (2013)
B. calyciflorus	CuO nanoparticles	Mortality	Manusadzianas et al. (2011).
B. calyciflorus	Metals and polycyclic aromatic hydrocarbons on fine air particles ($PM_{2.5}$)	Mortality	Verma et al. (2013)

(continued)

Table 9.1 (continued)

Species	Toxicants	Endpoints	Reference
B. koreanus	Benzo[a]pyrene	Population growth rate, cytochrome P450, glutathione s-transferase, GPx, MnSOD, CuZnSOD, catalase expression	Kim et al. (2013)
Plationus patulus	Paracetamol, diclofenac	Survival and reproduction	Sarma et al. (2014)
B. calyciflorus	5-Fluorouracil, capecitabine, cisplatin, doxorubicin, etoposide, imatinib	Mortality	Parrella et al. 2014
B. koreanus	Acetaminophen, atenolol, carbamazepine, oxytetracycline, sulfamethoxazole, trimethoprim	Survival, reproduction, permeability glycoprotein gene expression, acetylcholinesterase activity	Rhee et al. (2012)
Plationus patulus	Acetamidophenol, caffeine, fluoxetine, triclosan	Mortality, population growth rate	Martinez-Gomez et al. (2014)
Asplanchna brightwellii, B. calyciflorus	Dimethoate	Swimming	Chen et al. (2013)
Euchlanis dilatata	Cadmium, mercury, lead, methyl parathion	Phospholipase A2 enzyme activity	Arias-Almeida and Rico-Martinez (2011b)
Lecane quadridentata	Carbaryl, methyl parathion	Mortality, esterase activity, population growth rate	Perez-Legaspi et al. (2012)
Lecane inermis	Cu, Al, Fe, Zn, Sn, and Mn	Mortality	Klimek et al. (2013)
Philodina rapida	Lead	Reproduction	Esbaugh et al. (2012)
Philodina acuticornis	Endosulfan	Mortality	Allinson et al. (2011)
B. calyciflorus	Fenitrothion	Reproduction	Lv et al. (2010)
B. calyciflorus	Cu, Zn, Cd, Cr, Mn	Lifespan	Xu et al. (2014)
B. plicatilis	Eserine, pyrithione	Swimming	Garaventa et al. (2010)
B. plicatilis species complex	Crude oil, dispersant	Mortality, population growth rate, resting egg hatching	Rico-Martinez et al. (2013b)

period, post-reproductive period, net reproduction rate, generation time, and life expectancy at hatching of *B. calyciflorus*. E2 is a natural estrogen produced by mammals and humans and is a component of human contraceptives. It is excreted into natural water bodies, where it is found at concentrations as high as

200 ng/L. Depending on concentration, E2 shortened *B. calyciflorus* embryonic development and extended the juvenile, reproductive, and post-reproductive periods. Steroid hormones like E2 are thought to be used by rotifers to regulate reproduction (Snell 2011).

9.3 Bioconcentration: Lead

Lead (Pb) is a nonessential metal, ubiquitous in the environment, and has been listed by the US Environmental Protection Agency as one of the priority contaminants. Lead enters into aquatic systems by weathering and erosion or anthropogenic sources in the process of mining, industrial processing, and sewage disposal. Several studies showed that lead bioaccumulates and biomagnifies in animals through ingestion (Soto-Jiménez et al. 2011; Rubio-Franchini and Rico-Martínez 2011; Alvarado-Flores et al. 2012).

Bioaccumulation is usually expressed by the ratio of chemical concentration in body tissues relative to exposure concentration of the chemical, and the degree is the result of balance between the input rate and the rate of elimination (Alvarado-Flores et al. 2012). But bioaccumulation is typically not this simple, because metal accumulation does not only depend on the availability of the pollutant in the environment but also on a whole range of biological, chemical, and environmental factors. Soto-Jiménez et al. (2011) investigated the bioaccumulation of Pb on a four-level food chain: phytoplankton (*Tetraselmis suecica*), zooplankton (*Artemia franciscana*), shrimp (*Litopenaeus vannamei*), and a predatory fish (*Haemulon scudderi*). They showed that the level of Pb in phytoplankton substantially increased, with bioconcentration factors averaging 930–3630. However, there was a pronounced decrease in Pb concentration from phytoplankton to zooplankton and from zooplankton to shrimp tissues with a bioaccumulation factor of less than 1. Despite the decrease in the assimilation efficiency of Pb in phytoplankton and zooplankton, bioaccumulation was observed in the two predators, with Pb concentration higher in the fish than in the shrimp (bioaccumulation factor > 1.0). Pb concentrations in fish in the exposed group were two to three times higher than in the unexposed control. Rubio-Franchini and Rico-Martinez (2011) also showed that Pb can be biomagnified when the top predators are invertebrates. They found that freshwater rotifer *Asplanchna brightwellii* feeding on Pb-exposed prey (*Moina micrura*) have 13.1 times more Pb after 48-h exposure than feeding on unexposed prey, with a bioconcentration factor of 490. Using metal histochemistry and fluorescence techniques, Alvarado-Flores et al. (2012) clearly showed evidence of Pb bioconcentration in this rotifer. They also showed that Pb pellets are bioaccumulated in the mastax and vitellarium of *Brachionus calyciflorus* with a bioconcentration factor of 115 after 24-h exposure. Both studies of Rubio-Franchini and Rico-Martinez (2011) and Alvarado-Flores et al. (2012) showed that diet is the main route of exposure to Pb in this rotifer.

The toxicity of lead in rotifers is related to calcium ion and pH of the culture media (Esbaugh et al. 2012). Generally, rotifers are among the most resistant zooplankton to heavy metal contamination (Ciszewski et al. 2013). Arias-Almeida and Rico-Martínez (2011a, b) found that *Euchlanis dilatata* is quite tolerant to Pb with 48-h LC50 of 35.3 μg/l, whereas for *Philodina rapida*, depending on the origin of the test animals, EC50 for population growth ranges from 10.6 to 154.9 μg Pb/L (Esbaugh et al. 2012). For *B. calyciflorus*, Grosell et al. (2006) calculated an EC20 of 125 μg Pb/L for survival and 307 μg/L for population growth.

9.4 Toxicity of Nanoparticles

Because of their useful properties, nanoparticles are becoming more widely used in industrial materials and manufacturing (Klaine et al. 2008). They eventually appear as contaminants in aquatic environments, so it is important to estimate their environmental impacts. As key microinvertebrate grazers in freshwater and coastal marine environments, rotifers have been widely used as model animals in ecotoxicological studies. In one of the first studies of nanoparticle effects on aquatic food webs, Snell and Hicks (2009) showed that *Brachionus manjavacas* exposed to 0.3 μg/L of 37 nm diameter nanoparticles experienced 50% lower population growth rate. These particles were chemically inert, so their adverse effects were directly attributable to their size. Larger particles caused no reduction in population growth rate and remained confined to the gut, thus implicating nanoparticle size as a critical factor in the ability to penetrate the gut wall and enter tissues. An important ecological consequence of lower population growth rates is that less rotifer biomass is available to the many fish, shrimp, and crab larvae that begin their lives feeding on rotifers. If nanomaterials are to be used safely with minimal environmental impacts, care must be taken about the size and concentrations of particles released into aquatic environments (Handy et al. 2008).

In addition to their inhibition of rotifer reproduction, nanoparticles may bioaccumulate in rotifers as they do in other zooplankters like *Daphnia* (Hou et al. 2013) and be transferred to higher trophic levels. There is evidence of trophic transfer of nanoparticles from ciliates to rotifers (Holbrook et al. 2008). Other topics in nanoecotoxicology requiring investigation include information on the bioavailability of nanoparticles to aquatic organisms, mechanisms of cellular uptake, nanoparticle-biomolecule interactions in cells, the transformation of nanoparticles within cells, the mechanisms of cytotoxicity, and the sensitivity of specific ecosystem processes to nanoparticle exposure (Schirmer et al. 2013). This is the minimum knowledge required to perform predictive, reliable ecological risk assessments.

Titanium oxide nanoparticles illustrate the promise and potential hazards nanoparticles. TiO_2 are widely used because of their photocatalytic properties, generating reactive oxygen species (ROS) when illuminated with UV radiation (Bundschuh et al. 2011). Generation of ROS makes these compounds self-cleaning, which is useful for building materials like cement and asphalt and for abating air pollution from NOx. They are also used in cosmetics and sunscreens as stabilizers and to enhance the penetration of vitamins and antioxidants into the skin (Minetto et al. 2014). With so many industrial applications, TiO_2 will undoubtedly leach into surface waters where its biological effects are poorly understood.

Clément et al. (2013) examined the ecotoxicity of different TiO_2 nanoparticle concentrations, exposures, sizes, and crystalline structures to various taxonomic groups (cladocerans, rotifers, algae, plants). Using *Brachionus plicatilis*, they compared the toxicity of anatase (a hydrophilic form of TiO_2) nanoparticles from 15 to 44 nm diameter. They reported that 48 h exposure to 25 nm particles produced 50% immobility (EC50) in the test population at 10.4 mg/L and 32 nm particles at 267.3 mg/L. This large reduction in toxicity with slightly larger particles (32 vs 25 nm) suggests that there is still much to be learned about the mechanisms of toxicity of nanoparticles. In comparison, *Daphnia magna* exposed to 25 nm nanoparticles for 72 h produced an EC50 about one third of that of *B. plicatilis*. Both the *D. magna* and *B. plicatilis* tests estimated acute toxicity, but the rotifers were not fed prior to the test, unlike *Daphnia*. Moreover, the tests were conducted in 20 ppt saltwater for *B. plicatilis* and freshwater for *D. magna*, reflecting their different ecological requirements. Therefore conclusions about the relative sensitivity of these test animals should be regarded as preliminary.

The toxicity of copper oxide nanoparticles (CuO, mean size 30 nm) to the freshwater rotifer *Brachionus calyciflorus* was investigated by Manusadzianas et al. (2011). They showed that, in their experimental system, the majority of the toxicity was caused by the nanoparticles, not by ionic Cu^{2+}. *B. calyciflorus* was considerably more sensitive to the CuO nanoparticles (24 h LC50 0.24 mg/L) than the alga *Nitellopsis obtusa* (96 h LC50 4.3 mg/L) or the shrimp *Thamnocephalus platyurus* (24 h LC50 9.8 mg/L). More comparative studies are necessary to understand the differences in nanoparticle toxicity in fresh and saltwater, and the *B. plicatilis/B. calyciflorus* test system may be ideally suited for such experiments.

9.5 Toxicity of Fine Aerosols

Aerosols of fine particulate matter (<2.5 μm, $PM_{2.5}$) have been linked with several adverse environmental effects, including reduced atmospheric visibility, reduced nutrient cycling, and toxicity in vertebrates and invertebrates (Verma et al. 2013). These particles often contain metals and polycyclic aromatic hydrocarbons (PAHs) in high enough concentrations to elicit toxic responses in animals. Atmospheric deposition of these particles into water bodies can cause significant toxic loadings that can suppress animal population growth and compromise ecosystem function. Verma et al. (2013) examined the ecotoxicology of these particles using the rotifer *Brachionus calyciflorus*. They compared the acute toxicity of water and methanol extracts from these particles and consistently found the latter to be about eight times more toxic in eight samples taken from an urban Atlanta site over 2 months. Passing the methanol extracts over a C-18 column removed approximately 70% of the toxicity. These results suggest that the hydrophobic fraction of $PM_{2.5}$ extracts is responsible for most of the observed toxicity. This study is significant because it is one of the first to examine the toxicity of ambient $PM_{2.5}$ to invertebrates that are key components of aquatic food webs. Moreover, it demonstrates the facility of using *B. calyciflorus* for toxicity identification evaluations of small sample volumes.

The effect of the PAH benzo[a]pyrene on *Brachionus koreanus* was studied by Kim et al. (2013). Population growth rate of *B. koreanus* was significantly suppressed at 100 μg/L B[a]P, but not at 10 μg/L. They also found that B[a]P at 10 μg/L specifically induced expression of the cytochrome P450 proteins CYP2/3, but not other members of this gene family. These authors also reported that B[a]P exposure caused significant upregulation of several types of glutathione s-transferases as well as the antioxidant enzymes GPx, MnSOD, CuZnSOD, and catalase. B[a]P exposure also induced transcripts of high-molecular weight heat shock genes like hsp70, hsc70, hsp90α, and hsp90β, whereas transcripts of hsp21, hsp30, hsp40, and hsp60 decreased. Clearly, rotifers have molecular defenses to detoxify PAHs, and toxicity is manifested with exposures in the 10–100 μg/L range.

9.6 Toxicity of Pharmaceuticals

As pharmaceuticals are more utilized by humans, they are commonly appearing in the effluents of wastewater treatment plants (Fent et al. 2006). Their effects on the zooplankton of receiving waters are not well known, and rotifers have been important tools in describing zooplankton responses to these compounds. Sarma et al. (2014) quantified the chronic effects of the nonsteroidal anti-inflammatory pain-relieving drugs paracetamol and sodium diclofenac on the population dynamics of the rotifer *Plationus patulus* and the cladoceran *Moina macrocopa*. These drugs are widely consumed in industrialized countries and end up in wastewaters in fairly high concentrations. For example, wastewaters from the Tula Valley (State of Hidalgo, Mexico) contain concentrations of diclofenac at 2–5 μg/l. The rate of *P. patulus* population increase per day (r) was inversely related to the concentration of paracetamol or diclofenac in the medium over a concentration range of 1.6–32 mg/l. The r-values became negative when *P. patulus* was exposed to paracetamol at 32 or diclofenac at 25 mg/l. Significant suppression of rotifer population growth was observed at concentrations of 16 mg/l paracetamol and 6.25 mg/l diclofenac in 25 day exposures. It is important to characterize rotifer responses to a wide variety of pharmaceuticals so that we can anticipate when to be concerned about exposure concentrations in natural environments. It should be kept in mind, however, that toxicant sensitivity is not a fixed parameter, but varies with season, diet, and physiological condition. Toxicity thresholds estimated in well-fed laboratory populations in good conditions could well be higher than in natural populations that are often undernourished or otherwise stressed.

The environmental risks of exposure to six anticancer drugs used in chemotherapy, 5-fluorouracil, capecitabine, cisplatin, doxorubicin, etoposide, and imatinib, were estimated using the aquatic herbivores *Daphnia magna*, *Ceriodaphnia dubia*, *Brachionus calyciflorus*, and *Thamnocephalus platyurus* (Parrella et al. 2014). Acutely toxic effects occurred at mg/l concentrations, which are considerably higher than those predicted to occur in natural water bodies. The most toxic drugs were cisplatin and doxorubicin for most aquatic animals. For chronic endpoints, cisplatin and 5-fluorouracil showed the highest toxicity, inhibiting crustacean reproduction by 50% at μg/l concentrations. Rotifers were generally less susceptible to these pharmaceuticals than the crustaceans, but they were most sensitive to imatinib (LC50 = 3.82 mg/l). It is too early to draw conclusions about the environmental risk of these compounds because too little is known about their environmental fates, rates of biodegradation, and potential for bioconcentration.

Rhee et al. (2012) exposed *Brachionus koreanus* to six pharmaceuticals (acetaminophen, atenolol, carbamazepine, oxytetracycline, sulfamethoxazole, and trimethoprim) and recorded survival and reproduction in concentrations ranging from 10 to 100 μg/l. They also observed dose- and time-dependency effects on transcription of the Bk-P-gp gene, a permeability glycoprotein (P-glycoprotein). It is a drug efflux transporter of the ATP-binding cassette (ABC) transporter superfamily. Acetaminophen, carbamazepine, and oxytetracycline significantly induced P-gp gene expression at concentrations as low as 10 μg/l. All six compounds were inhibitory to *B. koreanus* population growth rate at concentrations of 100 μg/l. Rhee et al. (2013) also investigated the effects of these same compounds on the activity of acetylcholinesterase (AChE; EC 3.1.1.7) and found that this enzyme also was useful as a biomarker of pharmaceutical toxicity in rotifers. However, little is known about the concentrations of these compounds that are typically found in the environment, so the risk they pose to zooplankton is unknown.

The ecotoxicology of pharmaceutical and personal care products on *Plationus patulus* in an urbanized stretch of the Rio Grande river was investigated by Martinez-Gomez et al. (2014). Two-day exposure to acetamidophenol, caffeine, fluoxetine, and triclosan produced LC50s of 121, 419, 0.19, and 0.32 mg/l, respectively. Six-day exposure of *P. patulus* to acetamidophenol significantly inhibited rotifer population growth at concentrations of 15 mg/l. Population growth was inhibited at 200, 0.02, and 0.005 mg/l for caffeine, fluoxetine, and triclosan, respectively. Comparison of the Rio Grande population of *P. patulus* to a geographically isolated population from a remote site in Mexico revealed higher tolerance of acetamidophenol and caffeine in the former, suggesting adaptation to these compounds in urban environments.

9.7 Using Non-brachionid Species in Toxicity Assessment

Since the 1980s, most ecotoxicology studies using rotifers have employed species in the genus *Brachionus* (Snell and Janssen 1995). This is because brachionids are easily cultured, they are widespread and abundant, and they have a long history of use in ecological studies. Moreover, resting eggs are commercially available, eliminating the need to maintain stock cultures to provide animals for toxicity tests. Standardized acute toxicity tests utilizing *B. calyciflorus* and *B. plicatilis* have been published (ASTM 1991), as well as a standardized reproductive test to estimate chronic toxicity (Standard Methods 1998).

Other rotifer species are beginning to be used to investigate a variety of ecotoxicological problems like sediment toxicity, pollution in particular habitats, and the impact of toxicity on predator-prey relationships. These have been summarized by Rico-Martinez et al. (2013a). When performing ecological risk assessments in aquatic environments, it is advisable to estimate toxicity using several members of the food web. Rotifers are distinct from other grazing zooplankters in that they are not arthropods like cladocerans and copepods. The phylum Rotifera seems to be especially sensitive to certain classes of toxicants like androgens, detergents, and copper and not others like insecticides.

Rubio-Francini and Rico-Martinez (2011) observed bioaccumulation and bioconcentration in the rotifer *Asplanchna brightwellii* feeding on lead-contaminated *Moina micrura*. The *Asplanchna* bioaccumulation factor for lead was 123,684, and 490 for bioconcentration. *Asplanchna*-fed *Moina* exposed to lead had 13.3 times more lead than *Asplanchna* exposed directly to lead for 48 h, thus confirming lead biomagnification. They found similar bioconcentration of lead in *Brachionus calyciflorus* (Alvarado-Flores et al. 2012). The toxicity of other metals (Al, Cd, Fe, Zn) to *A. brightwellii* has been determined by Santos-Medrano and Rico-Martinez (Santos-Medrano and Rico-Martínez 2013). Chen et al. (2013) investigated the effects of the organophosphate pesticide dimethoate on the swimming behavior of *A. brightwellii*. Dimethoate exposure at 0.4–1.6 mg/l reduced the angular speeds of this rotifer to only 13–23% of that of controls. *A. brightwellii* swimming was more sensitive to dimethoate than *B. calyciflorus*.

The rotifer *Euchlanis dilatata* has been used in ecotoxicological studies because it is ecologically distinct from *Brachionus* species. Most *Brachionus* species are planktonic, spending the largest amount of their time swimming in the water column. *E. dilatata* in contrast spends more of its time attached to substrates and therefore is more likely to be exposed to toxicity in sediments (Arias-Almeida and Rico-Martinez 2011a). Arias-Almeida and Rico-Martinez (2011b) exposed *E. dilatata* to sublethal concentrations of three metals and two pesticides and measured inhibition of the in vivo activity of esterase and phospholipase A2 enzymes. They found both enzymes to be useful biomarkers of toxicant effect, with esterases most sensitive to cadmium and methyl parathion and phospholipases A2 more sensitive to mercury and lead. In acute toxicity tests, *E. dilatata* was generally more sensitive than *B. calyciflorus* or *Lecane quadridentata* (Arias-Almeida and Rico-Martinez 2011a).

The littoral rotifer *Lecane quadridentata* is becoming widely used in ecotoxicological studies because of its ecological relevance for assessing sediment toxicity (Pérez-Legaspi and Rico-Martínez 2001). Perez-Legaspi et al. (2012) compared the endpoints' 48-h mortality, 30-min in vivo inhibition of esterase activity, and 5-day inhibition of the population growth rate for their sensitivity to the pesticides carbaryl and methyl parathion. The population growth test had the lowest EC50s of

2.2 and 6.6 mg/l, respectively, for the two pesticides. Klimek et al. (2013a) examined the sensitivity to metal toxicity of *L. inermis* since it is being used to control activated sludge bulking by preventing overgrowth of filamentous bacteria in wastewater treatment plants. *L. inermis* is a prominent inhabitant of activated sludge, so it is important to know the metal concentrations that are likely to interfere with *L. inermis* bacterial grazing. They reported that *L. inermis* were most sensitive to Cu (LC50 = 25 μg/l), followed by Al, Fe, Zn, Sn, and Mn. This work demonstrates the upper concentrations of metals in activated sludge tolerable by *L. inermis* if they are to be effective in managing activated sludge bulking. In a second paper (Klimek et al. 2013b), these authors examined the sensitivity of *L. inermis* to aluminum salts. $AlCl_3$ and $Al_2(SO_4)_3$ are commonly used as chemical bulking agents for activated sludge, so it is important to know their toxicity to *L. inermis*. *L. inermis* were less sensitive to $AlCl_3$ (EC50 = 12 μg/l), especially at temperatures below 15 °C, so it may be feasible to simultaneously use both aluminum salts and *L. inermis* to stimulate bulking of activated sludge.

Bdelloid rotifers like *Philodina* also tend to crawl on the substrate, where they are more exposed to sediment toxicity. Esbaugh et al. (2012) examined *P. rapida* reproductive toxicity of lead spiked into five natural waters from North America. They found that Pb EC50s ranged from 10.6 to 154.6 μg/l and that Pb toxicity could be predicted by the calcium content and pH of the natural water at the collection sites. Some aspects of water chemistry do not alter toxicity. Allinson et al. (2011) reported that *P. acuticornis* in Australia were relatively insensitive to the acute toxicity of the pesticide endosulfan (EC50 = 1.75 mg/l) and that increasing salinity did not increase its toxicity.

9.8 Hormesis in Rotifers

Hormesis is the stimulatory effect of a compound at low concentrations when it is toxic at higher concentrations (Calabrese 2008). Typical dose-response curves are monotonically increasing adverse effects with increasing test compound concentration. As more compounds are tested, more cases of hormesis are being discovered, and they are increasingly recognized as adaptive responses (Forbes 2000). Below we describe several examples of hormetic responses by rotifers to a variety of toxicants.

Brachionus calyciflorus exposed to either the organochlorine pesticide dieldrin or 17β-estradiol at low concentrations (0.01–0.1 μg/L dieldrin, 0.001–0.1 μg/L 17β-estradiol) produced hormetic responses, shortening the duration of embryonic development and increasing the reproductive and post-reproductive periods, which lengthened lifespan (Huang et al. 2012). Exposure of *B. calyciflorus* to low concentrations of the organophosphate pesticide dimethoate increased population growth rate, but higher concentrations were toxic (Guo et al. 2012). *B. calyciflorus* exposed to 0.1 μg/L of the pesticide fenitrothion had longer reproductive periods and produced more offspring than controls, but higher concentrations (1000 μg/L) decreased offspring production (Lv et al. 2010). The lifespan of *B. calyciflorus* was extended by 30 and 36% by exposure to 0.01 and 0.1 mg Cd/L, respectively (Xu et al. 2014). Moreover, exposures to a mixture of Cu, Zn, Cd, Cr, and Mn at 0.01 or 0.1 mg/L concentrations also extended lifespan by 18%.

Low concentrations of these compounds also increased net reproduction rate. Exposure of *B. plicatilis* to 1 mg/L of the pesticide eserine or 0.001 mg/L of the antifouling compound Zn pyrithione produced a hormetic response, increasing swimming speed by 20–40% over a control value of 0.62 mm/s (Garaventa et al. 2010). Higher concentrations were toxic, markedly decreasing swimming speed. Several compounds are used as cryoprotectants to protect *B. plicatilis* cells from the damage of freezing (Prabu et al. 2014). These compounds include dimethyl sulfoxide (DMSO), glycerol, methanol, ethylene glycol, and propane diol and are typically protective at concentrations up to 10% and toxic at higher concentrations. The mechanisms by which they provide protection are not fully understood, but there is some evidence that they induce expression of some heat shock genes. *Brachionus manjavacas* exposed to 1 μM of the TOR gene inhibitor rapamycin or a JNK gene inhibitor responded with extended lifespan (Snell et al. 2014). Concentrations of these inhibitors greater than 10 μM were toxic, decreasing lifespan.

The fecundity of the rotifer *Brachionus urceus* exposed to UV-B radiation of 0.24 and 0.48 kJ/m^2 was 25% higher than controls, but fecundity was inhibited by higher UV exposures (0.96 and 1.20 kJ/m^2) (Wang et al. 2011). Clearly, hormesis is being more commonly found in rotifer ecotoxicology experiments. Rotifer scientists should be aware of this type of dose-response and be prepared to examine its mechanisms and adaptive significance when observed.

9.9 Screening Natural Compounds for Bioactivity

A major goal of aging research is to identify small, bioactive molecules capable of extending lifespan with minimum adverse side effects. One promising approach to identifying new molecules is to screen natural product libraries for bioactive compounds with antiaging effects. An effective experimental design is to use whole-animal screens to measure lifespan extension upon exposure to a diverse library of small molecules. This is likely to facilitate discovery of new biochemical pathways and mechanisms of action to target to slow aging. Because of our limited understanding of the mechanisms of aging, a simultaneous screen of multiple molecular targets is likely to be more productive than a more targeted approach.

Red algae continue to be a rich source of natural product discovery (Blunt et al. 2013). Snare et al. (2013) screened a library of Fijian red algae containing mixtures of methanol-extracted natural products for antiaging activity. They exposed the rotifer *Brachionus manjavacas* to 200 red algae extracts from 34 genera and tracked survival using life table analysis. Rotifer lifespan was increased 9–14% by doses of 2–10 μg/ml of eight of these 200 extracts. Three extracts were most effective at extending rotifer lifespan, including those from the red algae species *Acanthophora spicifera*, *Jania acutilobum*, and *Peyssonnelia* sp. Bioassay-guided fractionation led to semi-purified extracts composed primarily of lipids that were responsible for rotifer life extension. The life extending mixture from the red alga *A. spicifera* contained eicosanoic, octadecanoic, and hexadecanoic acids as well as several unidentified unsaturated fatty acids. The life extending effects of these small molecule mixtures were shown not to result from their direct antioxidant capacity. Studies like Snare et al. (2013) illustrate the potential of screening natural product libraries for producing strong candidates for the development of antiaging drugs.

Natural products also have been useful in aquaculture to manage rotifer populations. Essential oils extracted from the plants Origanum vulgare and O. marjorana were used as disinfectants for rotifer mass cultures (Stefanakis et al. 2013). Treatment of rotifer cultures with 10 mg/L of these oils for 4 h increased rotifer survival by about 50% and reduced numbers of presumptive *Vibrio* bacteria by about threefold.

9.10 Harmful Algae Blooms

Harmful algal blooms (HABs) are increasing in frequency, magnitude, and duration worldwide. Nearly 100 toxic or harmful species of dinoflagellates are known to cause severe damage to wild and aquacultured fish and shellfish. Dinoflagellates including several species of *Karenia*, *Heterocapsa*, and *Alexandrium* were reported to be toxic to rotifers. *Karenia* species which are known to be causative agents of neurotoxic shellfish poisoning in New Zealand (Chang and Ryan 2004) are reported to be lethal to *Brachionus plicatilis*, although different strains have different degrees of toxicity. For example, *Karenia mikimotoi* SUO-1 strain is lethal to *B. plicatilis* after 16-h exposure, while only 30% died at the same exposure time if they were exposed to *Karenia mikimotoi* FUK strain (Zou et al. 2010). Chang and Gall (2013) found that 50% of *Brachionus* sp. died after 20 and 60 min of exposure to single strength of lipophilic extract of *Karenia brevisulcata* and *Karenia concordia* extract, respectively, while it takes 20 h to kill 50% of the test rotifers when exposed to double strength *Karenia mikimotoi* extract. *Karenia brevisulcata* is known to produce high-molecular weight polyether toxins which are highly toxic to mice (Holland et al. 2012), while *K. mikimotoi* is known to produce low-molecular weight and hemolytic toxins (Neely and Campbell 2006), cytotoxic polyethers (Satake et al. 2002), and reactive oxygen species (Yamasaki et al. 2004). All of the above studies demonstrate that these dinoflagellates can inflict cell-mediated hemolytic damage to rotifers which leads to death.

Some paralytic shellfish poisoning (PSP) and non-PSP-producing strains of *Alexandrium* are also reported to be toxic to *B. plicatilis*. Wang et al. (2005) found that seven out of ten strains of *Alexandrium tamarense* are toxic to rotifers. In a study dosing rotifers with *Alexandrium* spp. at cell density of 2000 cells/ml, rotifers were killed with lethal time (LT50) ranging from 21 to 36 h. Three strains including a non-PSP-producing strain, *Alexandrium tamarense* (AT-6), and two PSP-producing strains (*Alexandrium lusitanicum*, *Alexandrium minutum*) did not significantly affect the population growth of *B. plicatilis*. In fact, addition of these strains as high as 2000 cells/ml increased population growth (Wang et al. 2005). This suggests that in some cases, *B. plicatilis* could be utilizing dinoflagellates as food.

9.11 Assessing Toxicity of Crude Oil

An accident at the Macondo oil well in the Gulf of Mexico in April 2010 spilled about 4.9 million barrels of crude oil into surrounding waters. The US EPA required the BP oil company to use the *Brachionus plicatilis* cyst-based acute toxicity test as part of the initial toxicity assessment (Rico-Martínez et al. 2013b). There was debate in the regulatory community about using dispersants on this spill and whether they actually reduced or amplified ecological impacts. Rico-Martinez et al. tested the toxicity of Macondo crude oil and the dispersant Corexit 9500A on several geographical strains of species in the *Brachionus plicatilis* species complex. Using water accommodated fractions (WAF) of oil, they found a range of LC50s from 2.5% WAF for the *B. plicatilis* Tokyo strain to 19.3% for the *Brachionus* sp. Veracruz strain. They argued that since the Veracruz strain is from the Gulf, perhaps it is adapted to the presence of small amounts of crude oil in its habitat. A reproductive endpoint was about fourfold more sensitive than acute toxicity, with a *B. manjavacas* LC50 of 11.0% WAF as compared to an EC50 of 2.6% WAF for reproduction and for cyst hatching. When the dispersant Corexit was added to the oil, toxicity increased up to 52 times more than the oil alone. Even though the oil-dispersant mixture is more toxic than oil alone, the dispersant is effective at reducing the amount of oil at the surface and redistributing it into the water column. Here, it is more likely to be toxic to zooplankters like rotifers and potentially impact fisheries that depend on the planktonic food web. Clearly a trade-off exists between dispersing oil from the surface, where it fouls beaches and contaminates seabirds and marine mammals, and dispersing it into the water column where it contaminates the pelagic food web. Coelho et al. (2013) criticized Rico-Martinez et al. (2013b) for not following standardized methods for assessing oil and dispersed oil toxicity. They also objected to using laboratory toxicity tests to extrapolate into the field for environmental assessment. Assessing the toxicity of complex mixtures like oil and dispersants has many pitfalls, and best practices are still being developed (Bejarano et al. 2014).

9.12 Future Perspectives

Rotifer immune response to toxicants seems like a rich area for investigation. Expression of genes involved in immune responses has proven useful as a bioindicator of animal exposure to xenobiotics. Shore and Ruvkun (2013) studied cytoprotective pathways that are activated by xenobiotics as part of innate immunity responses. They showed that detoxification and pathogen defense likely play a central role in the evolution of cytoprotective networks. Exposure to natural toxins from microbes throughout evolutionary history may have been a driving force behind the development of cytoprotection of essential activities like translation and metabolism. Disruption of these cell functions activates a pathogen and xenobiotic defense system, up-regulating detoxification genes like cyp-35B and protein chaperones, and pathogen-responsive genes like clec-60, irg-1, and F35H12.5. Detection of dysfunction triggers a regulatory cascade of responses, including signals that activate downstream effectors to coordinate systemic defenses through endocrine cues. Monitoring the activity of these genes could be useful in detecting xenobiotic exposures and predicting adverse effects.

Little is known about the innate immune capability of rotifers, but several studies have characterized immunity in mollusks, their lophotrochozoan cousins. In the mussel *Mytilus galloprovincialis*, hemocytes are the immune cells that carry out the molecular responses associated with immunity (Barcia and Ramos-Martínez 2011). These cells have the ability to respond to a variety of bacterial toxins by synthesizing catecholamines, activating phagocytosis, and increasing nitric oxide synthesis. Jenny et al. (2002) identified genes from the American oyster, *Crassostrea virginica*, which may be useful as biomarkers of xenobiotic exposure because they are involved in the stress response to environmental pollutants and infectious agents. These include an antimicrobial peptide, recognition molecules (lectin receptors), proteinases and proteinase inhibitors, and a novel metallothionein.

A similar approach may be fruitful using rotifers because transcriptome analyses (e.g., *Brachionus manjavacas* Transcriptome Shotgun Assembly project deposited at DDBJ/EMBL/GenBank, accession GARS00000000) have revealed the presence of similar innate immunity genes that could serve as bioindicators of xenobiotic exposure. These include interleukin enhancer binding factor, toll/interleukin-1 receptor, monocyte to macrophage differentiation protein, and a matrix metalloproteinase 2. A first step might be to investigate using qPCR how the expression of a gene like toll/interleukin-1 receptor is modulated by exposure to different classes of xenobiotics and potential pathogens. Another interesting line of investigation will be to examine whether xenobiotic exposure suppresses the innate immunity of rotifers, making them more susceptible to pathogens.

Thomas et al. (2011) described a parasitic oomycete fungus *Pythium* sp., infecting the rotifer *Asplanchna girodi* in a Georgia lake. They found that epidemics of this parasite typically occurred as three separate events in a single year. Prevalence at peak infection ranged from 29 to 41% and epidemics lasted from 17 to 56 days. These infections significantly reduced *A. girodi* fecundity, lifespan and population growth rate. Recent work on *Daphnia* suggests that *Pythium* sp. are important parasites of other zooplankters, but their effect on immune systems is unknown. Many questions remain like do *Pythium* infections render *Asplanchna* more susceptible to bacterial infections? Do pollutants compromise the *Asplanchna* immune response, making them more susceptible to *Pythium* infections? This area seems fertile for future investigations.

Acknowledgments We thank Roberto Rico-Martínez for comments that improved the manuscript.

References

Allinson G, Hagen T, Salzman S, Wightwick A, Nugegoda D (2011) Effect of increasing salinity on the acute toxicity of a commercial endosulfan formulation to the bdelloid rotifer *Philodina acuticornis odiosa*. Toxicol Environ Chem 93:722–728

Alvarado-Flores J, Montoya-Garcia MDR, Juárez JV, Rico-Martínez R (2009) Immunodetection of luteinizing hormone (LH), follicle-stimulating hormone (FSH), thyroid stimulating hormone (TSH) and prolactin (PRL) in *Brachionus calyciflorus* (Rotifera: Monogononta). Rev Biol Trop 57:1049–1058

Alvarado-Flores J, Rico-Martínez R, Ventura-Juárez J, Silva-Briano M, Rubio-Franchini I (2012) Bioconcentration and localization of lead in the freshwater rotifer *Brachionus calyciflorus* Pallas 1677 (Rotifera: Monogononta). Aquat Toxicol 109:127–132

American Society of Testing and Materials (1991) A standard practice for performing acute toxicity tests using rotifers in the genus *Brachionus*. 11.04, 1210–1216

Arias-Almeida JC, Rico-Martínez R (2011a) Toxicity of cadmium, lead, mercury and methyl parathion on *Euchlanis dilatata* Ehrenberg 1832 (Rotifera: Monogononta). Bull Environ Contam Toxicol 87:138–142

Arias-Almeida JC, Rico-Martínez R (2011b) Inhibition of two enzymatic systems in *Euchlanis dilatata* (Rotifera) as biomarkers of effect for metals and pesticides. Biomarkers 16:12–19

Barcia R, Ramos-Martínez JI (2011) Stress-based modulation of the immune response in molluscan hemocytes: a two-receptor model. Invertebr Surviv J 8:56–58

Bejarano AC, Clark JR, Coelho GM (2014) Issues and challenges with oil toxicity data and implications for their use in decision making: a quantitative review. Environ Toxicol Chem 33:732–742

Bundschuh M, Zubrod JP, Englert D, Seitz F, Rosenfeldt RR, Schulz R (2011) Effects of nano-TiO_2 in combination with ambient UV-irradiation on a leaf shredding amphipod. Chemosphere 85:1563–1567

Calabrese EJ (2008) Hormesis: why it is important to toxicology and toxicologists. Environ Toxicol Chem 27:1451–1474

Chang FH, Gall M (2013) Pigment compositions and toxic effects of three harmful *Karenia* species, *Karenia concordia*, *Karenia brevisulcata* and *Karenia mikimotoi* (Gymnodiniales, Dinophyceae), on rotifers and brine shrimps. Harmful Algae 27:113–120

Chang FH, Ryan KG (2004) *Karenia concordia* sp. nov. (Gymnodiniales, Dinophyceae), a new nonthecate dinoflagellate isolated from the New Zealand northeast coast during the 2002 harmful algal bloom events. Phycologia 43:552–562

Ciszewski D, Aleksander-Kwaterczak U, Pociecha A, Szarek-Gwiazda E, Waloszek A, Wilk-Woźniak E (2013) Small effects of a large sediment contamination with heavy metals on aquatic organisms in the vicinity of an abandoned lead and zinc mine. Environ Monit Assess 185:9825–9842

Clément L, Hurel C, Marmier N (2013) Toxicity of TiO_2 nanoparticles to cladocerans, algae, rotifers and plants – effects of size and crystalline structure. Chemosphere 90:1083–1090

Coelho G, Clark J, Aurand D (2013) Toxicity testing of dispersed oil requires adherence to standardized protocols to assess potential real world effects. Environ Pollut 177:185–188

Dahms H-U, Hagiwara A, Lee J-S (2011) Ecotoxicology, ecophysiology, and mechanistic studies with rotifers. Aquat Toxicol 101:1–12

Duffy TA, Iwanowicz LR, McCormick SD (2014) Comparative responses to endocrine disrupting compounds in early life stages of Atlantic salmon, *Salmo salar*. Aquat Toxicol 152:1–10

Esbaugh AJ, Brix KV, Mager EM, De Schamphelaere K, Grosell M (2012) Multi-linear regression analysis, preliminary biotic ligand modeling, and cross species comparison of the effects of water chemistry on chronic lead toxicity in invertebrates. Comp Biochem Physiol C 155:423–431

Fent K, Weston AA, Caminada D (2006) Ecotoxicology of human pharmaceuticals. Aquat Toxicol 76:122–159

Forbes VE (2000) Is hormesis an evolutionary expectation? Funct Ecol 14:12–24

Garaventa G, Gambardella C, Di Fino A, Pittore M, Faimali M (2010) Swimming speed alteration of *Artemia* sp. and *Brachionus plicatilis* as a sub-lethal behavioural end-point for ecotoxicological surveys. Ecotoxicology 19:512–519

García-García G, Sarma SSS, Núñez-Orti A-R, Nandini S (2014) Effects of the mixture of two endocrine disruptors (ethinylestradiol and levonorgestrel) on selected ecological endpoints of *Anuraeopsis fissa* and *Brachionus calyciflorus* (Rotifera). Int Rev Hydrobiol 99:166–172

Grosell M, Gerdes RM, Brix KV (2006) Chronic toxicity of lead to three freshwater invertebrates—*Brachionus calyciflorus, Chironomus tentans*, and *Lymnaea stagnalis*. Environ Toxicol Chem 25:97–104

Guo R, Ren X, Ren H (2012) Effects of dimethoate on rotifer *Brachionus calyciflorus* using multigeneration toxicity tests. J Environ Sci Health B Pestic, Food Contaminants, Agric Wastes 47:883–890

Handy RD, Owen R, Valsami-Jones E (2008) The ecotoxicology of nanoparticles and nanomaterials: current status, knowledge gaps, challenges, and future needs. Ecotoxicology 17:315–325

Holbrook RD, Murphy KE, Morrow JB, Cole KD (2008) Trophic transfer of nanoparticles in a simplified invertebrate food web. Nat Nanotechnol 3:352–355

Holland PT, Shi F, Satake M, Hamamoto Y, Ito E, Beuzenberg V, McNabb P, Munday R, Briggs L, Truman P, Gooneratne R, Edwards P, Pascal SM (2012) Novel toxins produced by the dinoflagellate *Karenia brevisulcata*. Harmful Algae 13:47–57

Hou W-C, Westerhoff P, Posner JD (2013) Biological accumulation of engineered nanomaterials: a review of current knowledge. Environ Sci: Process Impacts 15:103–122

Huang L, Xi Y, Zha C, Zhao L, Wen X (2012) Effects of dieldrin and 17β-estradiol on life history characteristics of freshwater rotifer *Brachionus calyciflorus* Pallas. J Freshw Ecol 27:381–392

Jenny MJ, Ringwood AH, Lacy ER, Lewitus AJ, Kempton JW, Gross PS, Warr GW, Chapman RW (2002) Potential indicators of stress response identified by expressed sequence tag analysis of hemocytes and embryos from the American oyster, *Crassostrea virginica*. Mar Biotechnol 4:81–93

Jones BL, VanLoozen J, Kim MH, Miles SJ, Dunham CM, Williams LD, Snell TW (2013) Stress granules from in *Brachionus manjavacas* (Rotifera) in response to a variety of stressors. Comp Biochem Physiol A 166:375–384

Kim R-O, Kim B-M, Jeong C-B, Nelson DR, Lee J-S, Rhee J-S (2013) Expression pattern of entire cytochrome P450 genes and response of defensomes in the benzo[a]pyrene-exposed Monogonont rotifer *Brachionus koreanus*. Environ Sci Technol 47:13804–13812

Klaine SJ, Alvarez PJJ, Batley GE, Fernandes TF, Handy RD, Lyon DY, Mahendra S, McLaughlin MJ, Lead JR (2008) Nanomaterials in the environment: behavior, fate, bioavailability and effects. Environ Toxicol Chem 27:1825–1851

Klimek B, Fialkowska E, Fyda J, Kocerba-Soroka W, Pajdak-Stos A, Sobczyk M (2013a) The toxicity of aluminum salts to *Lecane inermis* rotifers: are chemical and biological methods used to overcome activated sludge bulking mutually exclusive? Arch Environ Protect 39:127–138

Klimek B, Fialkowska E, Kocerba-Soroka W, Fyda J, Sobczyk M, Pajdak-Stos A (2013b) The toxicity of selected trace metals to *Lecane inermis* rotifers isolated from activated sludge. Bull Environ Contam Toxicol 91:330–333

Lv L, Snell TW, Yang J, Li S, Zhu W (2010) Effects of fenitrothion on life history parameters of the rotifer *Brachionus calyciflorus*. J Freshwat Ecol 25:589–598

Manusadzianas L, Caillet C, Fachetti L, Glyte B, Grigutyte R, Jurkoniene S, Karitonas R, Sadauskas K, Thomas F, Vitkus R, Ferard J-F (2011) Toxicity of copper oxide nanoparticles suspensions to aquatic biota. Environ Toxicol Chem 31(1):108–114

Martinez-Gomez DA, Baca S, Walsh EJ (2014) Lethal and sublethal effects of selected PPCPs on the freshwater rotifer, *Plationus patulus*. Environmental toxicology and chemistry, submitted

Minetto D, Libralto G, Ghirardini AV (2014) Ecotoxicity of engineered TiO_2 nanoparticles to saltwater organisms: an overview. Environ Int 66:18–27

Neely T, Campbell L (2006) A modified assay to determine hemolytic toxin variability among *Karenia* clones isolated from the Gulf of Mexico. Harmful Algae 5:592–598

Parrella A, Lavorgna M, Criscuolo E, Russo C, Fiumano V, Isidori M (2014) Acute and chronic toxicity of six anticancer drugs on rotifers and crustaceans. Chemosphere 115:59–66. http://dx.doi.org/10.1016/j.chemosphere.2014.01.013

Pérez-Legaspi IA, Rico-Martínez R (2001) Acute toxicity tests on three species of the genus *Lecane* (Rotifera: Monogononta). Hydrobiologia 446(447):375–381

Pérez-Legaspi IA, Quintanar JL, Rico-Martínez R (2012) Comparing toxicity endpoints on *Lecane quadridentata* (Rotifera: Monogononta) exposed to two anticholinesterases pesticides. Environ Toxicol 27:518–525

Prabu P, Rameshkumar N, Kayalvizhi N, Munuswamy N (2014) Effect of various cryoprotectants in the preservation of *Brachionus plicatilis* embryo and its role in expression profile of heat shock protein (HSP70). Int J Curr Microbiol App Sci 3:8–17

Rhee J-S, Jeong C-B, Kim B-M, Lee J-S (2012) P-glycoprotein (P-gp) in the monogonont rotifer, *Brachionus koreanus*: molecular characterization and expression in response to pharmaceuticals. Aquat Toxicol 114(115):104–118

Rhee J-S, Kim B-M, Jeong C-B, Park H-G, Leung KM-Y, Lee Y-M, Lee J-S (2013) Effect of pharmaceuticals exposure on acetylcholinesterase (AchE) activity and on the expression of AchE gene in the monogonont rotifer, *Brachionus koreanus*. Comp Biochem Physiol C 158:216–224

Rico-Martínez R, Pérez-Legaspi IA, Arias-Almeida JC, Santos-Medrano GE (2013a) Rotifers in ecotoxicology. In: Ferard JF, Blaise C (eds) Encyclopedia of aquatic toxicology. Springer-Verlag, Berlin/Heidelberg, pp 973–996

Rico-Martínez R, Snell TW, Shearer T (2013b) Synergistic toxicity of Macondo crude oil and dispersant Corexit 9500A® to the *Brachionus plicatilis* species complex (Rotifera). Environ Pollut 173:5–10

Rubio-Franchini I, Rico-Martínez R (2011) Evidence of lead biomagnification in invertebrate predators from laboratory and field experiments. Environ Pollut 159:1831–1835

Santos-Medrano GE, Rico-Martínez R (2013) Lethal effects of five metals on the freshwater rotifers *Asplanchna brightwelli* and *Brachionus calyciflorus*. Hydrobiologia 23:82–86

Sarma SSS, Gonzalez-Pérez BK, Moreno-Gutiérrez RM, Nandini S (2014) Effect of paracetamol and diclofenac on population growth of *Plationus patulus* and *Moina macrocopa*. J Environ Biol 35:119–126

Satake M, Shoji M, Oshima Y, Naoki H, Fujita T, Yasumoto T (2002) Gymnocin-a, a cytotoxic polyether from the notorious red tide dinoflagellate, *Gymnodinium mikimotoi*. Tetrahedron Lett 43:5829–5832

Schirmer K, Behra R, Sigg L, Suter MJ-F (2013) Ecotoxicological aspects of nanomaterials in the aquatic environment. In: Luther W, Zweck A (eds) Safety aspects of engineered nanomaterials. Pan Stanford Publishing Pte. Ltd. (www.panstanford.com) ISBN 978–981–4364-85-0

Shore DE, Ruvkun G (2013) A cytoprotective perspective on longevity regulation. Trends Cell Biol. https://doi.org/10.1016/j.tcb.2013.04.007

Snare DJ, Fields AM, Snell TW, Kubanek J (2013) Lifespan extension of rotifers by treatment with red algal extracts. Exp Gerontol 48:1420–1427

Snell TW (2011) A review of the molecular mechanisms of rotifer reproduction. Hydrobiologia 662:89–97

Snell TW, Hicks D (2009) Assessing toxicity of nanoparticles using *Brachionus manjavacas* (Rotifera). Environ Toxicol 26:146–152

Snell TW, Johnson RK, Rabeneck B, Zipperer C, Teat S (2014) Joint inhibition of TOR and JNK pathways interacts to extend the lifespan of *Brachionus manjavacas* (Rotifera). Exp Gerontol 52:55–69

Soto-Jiménez MF, Arellano-Fiore C, Rocha-Velarde R, Jara-Marini ME, Ruelas-Inzunza J, Páez-Osuna F (2011) Trophic transfer of lead through a model marine four-level food chain: *Tetraselmis suecica*, *Artemia franciscana*, *Litopenaeus vannamei*, and *Haemulon scudderi*. Arch Environ Contam Toxicol 61:280–291

Standard Methods for the Examination of Water and Wastewater (1998) Estimating chronic toxicity using rotifers. 20(8420)8-62–8-65

Stefanakis MK, Anastasopoulos E, Katerinopoulos HE, Makridis P (2013) Use of essential oils extracted from three *Origanum* species for disinfection of cultured rotifers (*Brachionus plicatilis*). Aquac Res. doi:10.1111/are.12137

Stout EP, La Clair JJ, Snell TW, Shearer TL, Kubanek J (2010) Conservation of progesterone hormone function in invertebrate reproduction. Proc Natl Acad Sci 107:11859–11864

Thomas SH, Housely JM, Reynolds AN, Penczykowski RM, Kenline KH, Hardegree N, Schmidt S, Duffy MA (2011) The ecology and phylogeny of oomycete infections in *Asplanchna* rotifers. Freshw Biol 56:384–394

Verma V, Rico-Martínez R, Kotra N, Rennolds C, King L, Liu J, Snell TW, Weber RJ (2013) Estimating the toxicity of ambient fine aerosols using freshwater rotifer *Brachionus calyciflorus* (Rotifera: Monogononta). Env Pollution 182:379–384

Vos JG, Dybing E, Greim HA, Ladefoged O, Lambre C, Tarazona JV, Brandt I, Vethaak AD (2000) Health effects of endocrine-disrupting chemicals on wildlife, with special reference to the European situation. Crit Rev Toxicol 30:71–133

Wang L, Yan T, Yu R, Zhou M (2005) Experimental study on the impact of dinoflagellate *Alexandrium* species on populations of the rotifer *Brachionus plicatilis*. Harmful Algae 4:371–382

Wang J, Feng L, Tang X (2011) Effect of UV-B radiation on the population dynamics of the rotifer *Brachionus urceus*. Acta Oceanol Sin 30:113–119

Xu X, Xi Y, Huang L, Xiang X (2014) The life-table demographic response of freshwater rotifer *Brachionus calyciflorus* to multi-metal (Cu, Zn, Cd, Cr, and Mn) mixture interaction. Bull Environ Contam Toxicol. doi:10.1007/s00128-014-1281-y

Yamasaki Y, Kim D, Matsuyama Y, Oda T, Honjo T (2004) Production of superoxide anion and hydrogen peroxide by the red tide dinoflagellate *Karenia mikimotoi*. J Biosci Bioeng 97:212–215

Zhao L-L, Xi Y-L, Huang L, Zha C-W (2009) Effects of three phthalate esters on the life-table demography of freshwater rotifer *Brachionus calyciflorus*. Aquat Ecol 43:395–402

Zou Y, Yamasaki Y, Matsuyama Y, Yamaguchi K, Honjo T, Oda T (2010) Possible involvement of hemolytic activity in the contact-dependent lethal effects of the dinoflagellate *Karenia mikimotoi* on the rotifer *Brachionus plicatilis*. Harmful Algae 9:367–373

Chapter 10
Rotifers in Ecotoxicology

Eun-Ji Won, Jeonghoon Han, Duck-Hyun Kim, Hans-U. Dahms, and Jae-Seong Lee

Abstract Rotifers are widespread and abundant in aquatic ecosystems and are valuable live feeds in aquaculture and fisheries. They are particularly useful in evaluating full life cycle and population-level effects in response to pollutant exposures. Rotifers have, therefore, been considered as a promising species in aquatic ecotoxicology. The importance of using rotifers in aquatic ecotoxicology was constantly reviewed (Snell and Janssen Hydrobiologia 313/314:231–247, 1995; Dahms et al. Aquat Toxicol 101:1–12, 2011). This chapter will focus on discussing, with diverse examples, why rotifers became promising model species in ecotoxicology, taking into account their biological, physiological, and genomic information. Ecotoxicogenomic approaches using RNA-seq and genomic approaches will be demonstrated. In particular, the monogonont rotifer *Brachionus* sp. will be introduced as a target taxon for ecotoxicogenomic studies with substantial genetic information.

Abbreviation

AhR	Aryl hydrocarbon receptor
AOPs	Adverse outcome pathways
B[*a*]P	Benzo[*a*]pyrene
CYP	Cytochrome P450
ERA	Environmental risk assessment
EST	Expressed sequence tags
GPx	Glutathione peroxidase

E.-J. Won • J. Han • D.-H. Kim • J.-S. Lee (✉)
Department of Biological Science, College of Science, Sungkyunkwan University, Suwon 16419, South Korea
e-mail: jslee2@skku.edu

H.-U. Dahms
Department of Biomedical Science and Environmental Biology, Kaohsiung Medical University, Kaohsiung, 80708 Taiwan, Republic of China

Department of Marine Biotechnology and Resources, National Sun Yat-sen University, No. 70, Lienhai Road, Kaohsiung, 80424 Taiwan, Republic of China

A. Hagiwara, T. Yoshinaga (eds.), *Rotifers*, Fisheries Science Series,
DOI 10.1007/978-981-10-5635-2_10

GR	Glutathione reductase
GSH	Glutathione
GST	Glutathione *S*-transferase
Hsps	Heat shock proteins
IR	Ionizing radiation
LC	Lethal concentration
LD	Lethal dose
MAPK	Mitogen-activated protein kinase
MIEs	Molecular initiating events
MWCNTs	Multi-walled carbon nanotubes
ROS	Reactive oxygen species
SOD	Superoxide dismutase
SULTs	Sulfotransferases
TPT	Triphenyltin

10.1 Introduction

Rotifers are promising model species in the evaluation of environmental pollution, as they have a central role in aquatic ecosystems and they are easy to cultivate, providing quantitative endpoints with standardized protocols. Such endpoints include whole life span parameters including hatching, growth, mortality, and reproduction for ecotoxicological studies. Also a full life-cycle toxicity test was developed using their resting eggs for ecological assessments (Preston and Snell 2001). In a recent review on molecular approaches with rotifers, several advantages of rotifers in experimental studies were summarized (Table 10.1; Dahms et al. 2011). Also rotifers using cilia for whirling water movements to trap food maximize their uptakes of any harmful materials from the water phase through the digestive tract during intake of organic particles. In addition, the body with transparent cuticles from the mouthpart to gut allows to evaluate organ effects of pollutants by viewing alterations of the internal anatomy (Juchelka and Snell 1994; Snell and Hicks 2011). For example, microplastics and nanoparticles were optically measured in rotifers under the microscope after oral uptake in several toxicity tests (Juchelka and Snell 1994; Snell and Hicks 2011). Easy maintenance of cultures under laboratory conditions allows rotifers to be both useful feed organisms in aquaculture and model species in ecotoxicological studies (Austad 2009; Dahms et al. 2011). In the 1970s, the rotifer *Philodina roseola* was firstly noticed in toxicological testing to assess the toxicity of chromate and arsenate (Schaefer and Pipes 1973). Since then, many studies have continued this line of research using other rotifers as model species.

The phylum Rotifera comprises three classes: Seisonida, Monogononta, and Bdelloidea. Among them, Monogononta is the most dominant class comprising of more than 77% of rotifer species (Segers 2007). The monogonont rotifer genus *Brachionus* is widely distributed along coastlines and plays important roles in

Table 10.1 Advantages of rotifers in experimental studies

Small size	Small size allows the use of small volumes for rearing and experimentation
First consumer in ecosystem	Main linker of microbial food webs
Simple organization	Simple organization allows the analysis of complex processes through straightforward approaches
Transparency (body)	Easy to observe
Life cycles	Short life cycles allow the study of multigeneration effects in reasonable periods of time
Parthenogenesis	Rotifers commonly reproduce parthenogenetically – this provides genetic homogeneity, fast population growth, and high densities, since all individuals are females and can contribute to the next generation
Genetic homogeneity	Offspring resulting from ameiotic parthenogenesis is genetically identical
Variability	Variable responses can be pronounced in between species: e.g., interspecific differences in response to toxicants can be substantial
Resistance	Dormant resting stages provide rotifers with increased resistance to a wide range of environmental stressors
Resting eggs	Resting eggs in the Monogononta can be stored and hatched when necessary for experimentation

Some issues are modified after Dahms et al. (2011)

aquatic ecosystems as a main linker in microbe-based food webs (Arndt 1993; Dahms et al. 2011; Rego et al. 2015). In particular, *B. plicatilis* has been well studied with regard to its population dynamics (Ortells et al. 2000, 2003), speciation (Gomez et al. 2002; Suatoni et al. 2006), evolution of sexual reproduction (Stelzer and Snell 2003; Snell et al. 2006), and ecotoxicology (Snell and Janssen 1995; Snell et al. 2003). Specific traits including morphology, reproduction, and mating behavior of rotifers have also been studied in *Brachionus* sp. (Hagiwara et al. 1995; Sha et al. 2015). Recently, next-generation sequencing (NGS) techniques allow principle biological studies toward further diagnosis using genomic information (Wang et al. 2015). For example, the database of over 2300 expressed sequence tags (EST) corresponding to more than 450 transcripts was constructed by Suga et al. (2007). The molecular characteristics associated with resting eggs and survival during dormancy and hatching were studied in *B. plicatilis* by comparing relevant ESTs (47,926 ESTs were assembled into ~18,000 putative transcripts) (Denekamp et al. 2009). An Illumina RNA-seq study also released a large number of reads from the resting eggs of *B. plicatilis* (Clark et al. 2012). RNA-seq information of the intertidal rotifer *B. koreanus* was released recently (Lee et al. 2015a), and this species can be one of the emerging model species in ecotoxicology (Dahms et al. 2011). In this chapter, we mainly summarize research on individual and cellular/subcellular endpoints that were made particularly with representatives of *Brachionus*. We will particularly highlight here research areas where rotifers were used as well as future perspectives in the field of ecotoxicology.

10.2 Physiological Effects of Rotifers on Ecotoxicology: Individual Endpoints

10.2.1 Mortality

Rotifers have been used as prominent indicator species for studying the adverse effects of pollutants and for using biomonitoring in the aquatic environment and ecosystem with several advantages to conduct experimental studies (Gannon and Stemberger 1978; Ortells et al. 2000; Gutkowska et al. 2013) (Table 10.1). Rotifers contribute to the nutrient cycling in aquatic ecosystem due to their fast population growth and rapid turnover rates by having a short life span (less than 3–4 days). This rapid turnover allows them to respond more quickly to environmental changes, leading to be a sensitive indicator of changing water quality relative to other aquatic invertebrates such as crustacean plankton (Gannon and Stemberger 1978).

In toxicological assessments, physiological indexes require simple and inexpensive toxicity testing. Among them, the lethal dose (LD) and lethal concentration (LC) are essential and easy-to-obtain endpoints for ecotoxicological assessments. In acute toxicity tests using rotifers, a standard test was carried out in a 24-h survey with neonates (less than 12 h post hatching) to obtain values for 50% survival at a lethal concentration after 24 h as a short-term life-cycle test. Until 1978, only mortality data had been considered as an index for evaluating toxicity in rotifers. In the late 1980s, a cyst-based acute toxicity test was standardized for *B. plicatilis* and *B. rubens* (Snell and Persoone 1989a, b). The provision of rotifer resting eggs (cysts) by storage at dry conditions facilitated toxicity testing (Hagiwara 1994; Kim and Hagiwara 2009). The authors provided a kit ROTOXKITS™ for acute and short chronic tests using *B. calyciflorus* (MicroBioTests Inc., Kleimoer, Belgium).

In *Brachionus* species, the lethal effect of free chloride was firstly reported in *B. plicatilis* (Capuzzo 1979). Using *B. calyciflorus* and *B. plicatilis*, several acute toxicity datasets were collated and reported (Snell and Janssen 1995). After this review from 1995, data on lethal effects have been newly compiled here for *Brachionus* spp. (Table 10.2). Particularly, in *B. koreanus*, lethal effects were recently published to several stressors, such as in response to UV radiation (Kim et al. 2011), gamma radiation (Han et al. 2014), triphenyltin (TPT) (Yi et al. 2016), and several toxicants, and challenging environmental factors (e.g., copper, cadmium, benzo[*a*]pyrene, hydrogen peroxide, water-accommodated fraction of crude oil, pharmaceuticals, salinity, and temperature).

Interestingly, the values of LC50 or LD50 vary widely with the toxic substances applied. In *B. plicatilis*, acute toxicity testing for 11 different organic phosphorous compounds (diazinon, parathion-methyl, parathion, phorate, malathion, azinphos-ethyl, azinphos-methyl, dimethoate, chlorpyrifos, omethoate, fonofos) were conducted and revealed lethal sensitivities across two orders of magnitude (1.7 mg/L of chlorpyrifos and 295 mg/L of omethoate) (Guzzella et al. 1997). Also different species showed different susceptibilities for toxic substances even though the same protocol was applied. For example, in the case of copper exposure of three

Table 10.2 Integrated dataset of acute toxicity studies with rotifer *Brachionus* sp. during the last 20 years (from 1995 to 2015)

Species	Chemicals or environmental stress	Additional effect (exposure condition)	LC50/LD50 (95% CI)	Remarks	Reference
B. koreanus	UV radiation	(25 °C, 15°/$_{oo}$, 100 μW/cm^2)	24.6KJ/m^2 (23.8–25.4)		Kim et al. (2011)
	Copper	(25 °C, 15°/$_{oo}$)	1.2 mg/L (1.07–1.23)		Han et al. (2013)
	Tributyltin	(25 °C, 15°/$_{oo}$)	29.6 μg/L (27.2–32.1)		Yi et al. (2016)
	Gamma radiation	(25 °C, 7.81 gy/min, 15°/$_{oo}$)	2.94 kgy (2.73–3.17)		Unpublished data
B. plicatilis	PCP		1.93 mg/L		Moffat and Snell (1995)
	TBT		0.011 mg/L		
	Copper		0.063 mg/L		
	Mercury		0.61 mg/L		
	Zinc		<4.8 mg/L		
	Ca(OCl)2		0.56 mg/L		
	Copper	+ effects of salinity, pH, and DOM (at 24.7–25.4 °C)	38.5~, >442 μg/L		Arnold et al. (2010)
	Copper	Depending on feeding state	80.9 μg/L (unfed) <133.4 μg/L (fed)	*B. plicatilis* L strain	Arnold et al. (2011)
B. manjavacas	Crude oil + dispersant				Rico-Martínez et al. (2013)
B. plicatilis sensu stricto					
B. rotundiformis					
Brachionus sp.					

(continued)

Table 10.2 (continued)

Species	Chemicals or environmental stress	Additional effect (exposure condition)	LC50/LD50 (95% CI)	Remarks	Reference
B. plicatilis	OP (diazinon)		28 mg/L		Guzzella et al. (1997)
	OP (parathion-methyl)		>67 mg/L		
	OP (parathion)		>36 mg/L		
	OP (phorate)		>50 mg/L		
	OP (malathion)		74 mg/L		
	OP (azinphos-ethyl)		>5.2 mg/L		
	OP (azinphos-methyl)		85 mg/L		
	OP (dimethoate)		244 mg/L		
	OP (chlorpyrifos)		1.7 mg/L		
	OP (omethoate)		295 mg/L		
	OP (fonofos)		8.8 mg/L		
	Diazinon		88.39 μM (74.58–102.51)		Marcial et al. (2005)
	Fenitrothion		229.76 μM (219.30–240.22)		
	Isoprothiolane		220.74 μM (215.92–225.56)		
	Methoprene		100.81 μM (88.89–112.40)		

B. calyciflorus	Pentachlorophenol	+ UV radiation	83–210 μg/L	UV enhance chemical toxicity	Preston et al. (1999b)
	Mercury		51–64 μg/L		
	Copper	Depending on food density	0.19–0.62 mg/L		Sarma et al. (2000)
	Cadmium				
	Mercury				
	Pentachlorophenol		1200 μg/L		Preston et al. (2001)
	Copper		26 μg/L		
	AB (erythromycin)	25 °C, dark condition	0.94 mg/L (0.93–1.41)		Isidori et al. (2005)
	AB (oxytetracycline)		1.87 (1.19–2.96)		
	AB (sulfamethoxazole)		9.63 (7.00–13.25)		
	AB (ofloxacin)		0.53 (0.34–0.82)		
	AB (lincomycin)		0.68 (0.38–1.22)		
	AB (clarithromycin)		12.21 (10.43–14.72)		
	PFOS		61.8 mg/L		Zhang et al. (2013)
	PFOA		150.0 mg/L		
B. angularis	Methyl parathion	+ effect of food density	0.064–6.52 mg/L		Gama-Flores et al. (2004)

DOM dissolved organic matter, *OP* organic phosphorus, *AB* antibiotics

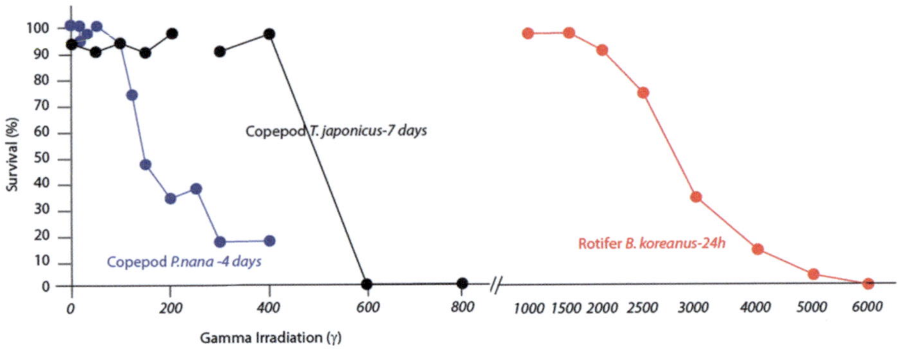

Fig. 10.1 Dose–response curves based on the survival rate of the copepods *Tigriopus japonicus* and *Paracyclopina nana* and the rotifer *B. koreanus* in response to gamma radiation (The results are adopted from Han et al. 2014; Won and Lee 2014; Han et al. unpublished)

Brachionus spp. (*B. plicatilis*, *B. koreanus*, and *B. calyciflorus*), *B. koreanus* showed a large tolerance capacity to mitigate oxidative stress induced by copper (LC50 = 1.2 mg/L; having 40 times tolerance than *B. calyciflorus*) (Table 10.2). Previously, the lethality in rotifer was lower than one order of magnitude compared to fathead minnow in some cases (Snell et al. 1991a). In other studies, however, scientists have concluded that the rotifer *Brachionus* sp. has higher susceptibility in acute tests of several compounds (Snell et al. 1991a, b; Persoone et al. 1993). Particularly, about 62% of the cases conducted with the freshwater rotifer *B. calyciflorus* with cyst-based acute toxicity tests were more sensitive than the conventional testing with the water flea *D. magna* (Persoone et al. 1993). Interestingly, the toxicity values of rotifers in response to gamma radiation are remarkably high compared to other small aquatic invertebrates such as copepods *Paracyclopina nana* and *Tigriopus japonicus* (Fig. 10.1). The LD50-24 h and LD50-96 h values for *B. koreanus* were approximately 2900 and 2300 Gy, respectively (Fig. 10.1, unpublished data). Also, in the bdelloid rotifer *Adineta vaga*, a great tolerance to ionizing radiation (IR) was observed with an unusually effective system of antioxidant protection (Krisko et al. 2012). Regarding the general high resistance of rotifers to gamma radiation, a high genetic homogeneity due to parthenogenetic reproduction is likely the reason of the high resistance to IR. Here, the active DNA repair system is able to use homologous chromosome templates that result from parthenogenetically cloned diploid eggs (Krisko et al. 2012). Different sensitivities do not only indicate differences of sensitivity but are also indicating different suitabilities of toxic tests (Persoone et al. 1993). In fact, the susceptibility of rotifers in response to each compound or environmental stress provides valuable information to fill the knowledge gaps in ecotoxicological studies with rotifers. Taken together, the variability of LC50 and LD50 values is wide in rotifers, showing a wide range of sensitivity in response to chemicals and other stressors that allow to use rotifers as one component of a battery system within ecotoxicological studies such as together with other organisms as shown in Fig. 10.1.

10.2.2 Growth Retardation and Reproductive Impairment

The population dynamics of experimental target organism is a good parameter to evaluate the impacts of environmental risks, as it can be regulated by environmental factors. For example, the measurement of growth or reproduction has been studied for a long time in aquatic ecology and ecotoxicology, as growth and reproduction are relatively straightforward parameters in aquatic biota. In rotifers, population dynamics is a promising endpoint of growth and reproduction that is negatively affected by environmental pollutants and stressors (Halbach et al. 1981; Halbach 1984). Particularly, the parthenogenesis (asexual reproduction) of rotifers facilitates quantitative measurements using life table techniques of isolated females and/or the log phase of population development (Snell and Janssen 1995). Thus, rotifers have been considered as suitable candidates and model species studying population dynamics (Halbach 1984).

Rotifer population dynamics was used in ecotoxicology as an endpoint for evaluating stress conditions in a lab-based experiment using *B. rubens* and *B. calyciflorus* (Halbach 1984). For measuring rotifer population dynamics, growth rate (*r*), and *K* (carrying capacity), the frequency and amplitude of population oscillations were used as life history endpoints. Also population growth using a two-day life-cycle test was used in *B. calyciflorus* as an endpoint for a chronic toxicity test (Snell and Moffat 1992). In several studies, the intrinsic rate of natural increase "*r*" or the number of rotifers substituted for population dynamics was a sensitive, rapid, and inexpensive measure of toxicity. This parameter was dose dependently reduced in response to several toxicants and stress exposure in several *Brachionus* species (Kim et al. 2011, 2014a, b; Han et al. 2014). For example, In *B. calyciflorus*, reproductive impairment was present within 3 days in response to ethinylestradiol, nonylphenol, and testosterone (Radix et al. 2002). In *B. plicatilis*, significant differences in *r* in response to light wavelengths and intensities were shown over phototaxic behavior, as photokinesis reduces population growth by increasing energy usage by elevating swimming speed (Kim et al. 2014b). In *B. koreanus*, UV-B and gamma irradiation altered life table parameters, particularly reproductive success and growth rate (Kim et al. 2011; Han et al. 2014). At a low dose of UV-B radiation (2 kJ/m^2, 100 μW/m^2), growth retardation was observed but was less than the NOEC value. In gamma-irradiated *B. koreanus*, the ability to recover gamma radiation-induced damages on fecundity was estimated after examining two generations with a sigmoidal growth curve and log–log plots of the exponential growth phase of each irradiated group (Fig. 10.2).

Egg production or hatching rate is also easily measured as reproduction success in rotifers. For example, the toxicity of dispersed oil from a Gulf of Mexico oil spill was examined in the rotifer *B. plicatilis* (Rico-Martínez et al. 2013). In five strains of the rotifer *B. plicatilis*, the toxicity of oil from the Deepwater Horizon spill, dispersant (Corexit®), and its mixture showed a fast response time and sensitivity to toxicants. The oil-dispersant mixture was much (>50×) more toxic than oil or the dispersant alone, causing an inhibition of egg hatching by 50%. As for the resting

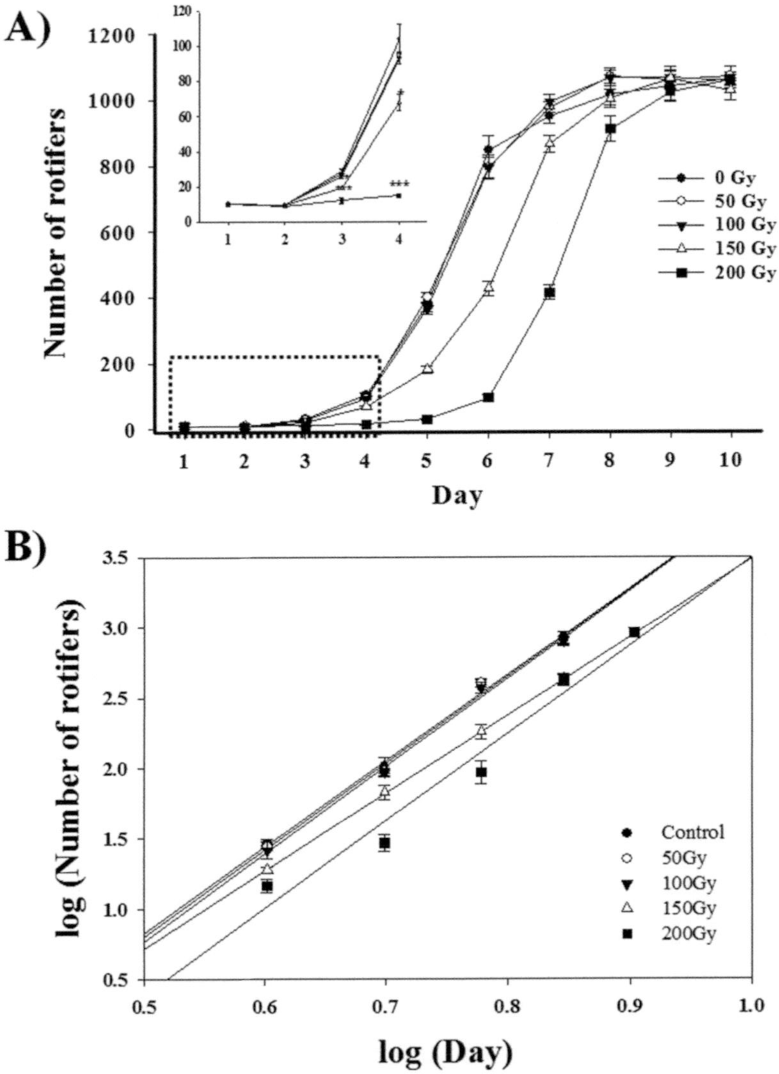

Fig. 10.2 Effect of gamma radiation on the fecundity of *B. koreanus*: (**a**) the population growth rate in response to gamma irradiation (0, 50, 100, 150, and 200 Gy) for 10 days and (**b**) log–log plots from the exponential growth phase of each group (Adopted from Han et al. 2014)

egg hatchability, the hatching rate of resting eggs showed consistent sensitivity in response to pesticide exposure (e.g., diazinon, fenitrothion, methoprene, and isoprothiolane) at approximately 2–40 times lower doses compared to EC50 values of growth rate, fertilization, and resting egg production (Marcial et al. 2005).

Table 10.3 Advantages and disadvantages of behavioral endpoints in rotifers for ecotoxicological studies

Advantages	Rapid (~minute)
	Automated data collection (in case using video microscopy and computer)
Disadvantages	High cost of video motion tracking system
	Some behavioral responses are temporary (often with unclear results)

10.2.3 Behavioral Responses in Toxicity Studies

The general movements including feeding behavior, predator avoidance, reproductive behavior, and social behavior are used as indices of physiological alterations in aquatic ecosystems, as behavior is closely linking between physiological and ecological processes. Behaviors particularly provide sensitive and precise parameters in response to stressors (e.g., environmental toxicants) and are useful to evaluate ecological effects at the population and community level. Previously, technical methods to measure the behavior of rotifers at toxic conditions have been well described (reviewed in Snell and Janssen 1995). In particular, advantages and disadvantages of behavioral endpoints (Table 10.3) emphasized that the behavioral endpoint is an attractive index when it is directly related to widely accepted adverse effects (e.g., reduced survival or reproduction rate).

Alteration of locomotion is very common in response to a variety of environmental stressors. In ecotoxicology, the locomotion of bivalves and crabs is closely related to their movements in response to stress conditions. In the juvenile clam *Macomona liliana* exposed to copper, they showed a slow rate of burial in response to an increase of copper concentration (Roper and Hickey 1993). Also, in the isopod *Saduria entomon*, reduced burrowing was a result of avoidance in metal contaminated sediments (10 μg/g dry sediment copper, 35 μg/g cadmium, or 200 μg/g iron) (Pynnönen 1996). However, in rotifers, a reliable measurement of locomotion is limited due to their small sizes (150–250 μm) and their planktonic characteristics. For example, in the rotifer *B. plicatilis*, the migration pattern measured by their movement at different light exposures showed that phototactic responses were not affected by light intensity directly. Instead they were affected by other factors (e.g., by cladocerans that are the main competitors and also predators of rotifers in nature) (Kim et al. 2014b).

Feeding behavior is also an interesting endpoint for ecotoxicological studies. It can be easily quantified by measuring the clearance of microalgal concentrations over a period of feeding (Ferrando et al. 1993). ^{13}C-labeled green algae have also been used to measure feeding rates (Verschoor et al. 2005). These approaches have been employed in ecology, as feeding behavior is directly associated with growth and reproduction. For example, ingestion rates were reduced in a dose-dependent manner over increasing toxicant concentrations in the rotifer *B. calyciflorus* in response to four different pollutants including 3,4, dichloroaniline, lindane, pentachlorophenol, and copper (Ferrando et al. 1993). Thus, in ecotoxicology, feeding behavior is a potential biomarker with a different perspective. The ability of prey to avoid predators is often impaired in response to contaminants, resulting in the reduction of survival and population size in the ecosystem at the presence of predators. In *B. calyciflorus* in response to the combined exposure of sublethal pentachlorophenol and the predator *Asplanchna girodi*, swimming speed of prey was enhanced by up to 30%, but the predation risk was increased up to 50% (Preston et al. 1999a). In conclusion, rotifer plays a critical role in the aquatic food web and organic cycles; thus, risk assessment using adverse outcomes observed at the individual level can supply important information for a systematic interpretation of ecotoxicological endpoints.

10.3 Biochemical Endpoints of Rotifers for Ecotoxicological Studies: Antioxidant and Detoxification Mechanisms of Enzymatic Activities

Bioassays with rotifer are commonly conducted in aquatic ecotoxicology since the 1970s. Of them, a wide variety of bioassays have been developed based on individual endpoints such as growth rate, ingestion rate, and mortality in rotifers as described in the previous section. Recently, however, the measurement of antioxidant enzymatic activities is considered as a promising endpoint. In aquatic vertebrates, biomarkers have been widely applied over time for detecting environmental contamination (Stegman 1978). Recently, several approaches including stress proteins and in vivo enzymatic activities have been conducted with aquatic invertebrates with various indices for ecotoxicology.

As a first report using biomarkers for assessing aquatic toxicity in rotifers, the increased expression of stress protein SP58, immunologically cross-reactive with the 65,000 Da heat shock protein, was measured in *B. plicatilis* in response to copper and tributyltin but was not correlated in response to aluminum chloride, mercury chloride, pentachlorophenol, sodium arsenite, sodium azide, sodium dodecyl sulfate, or zinc chloride (Cochrane et al. 1991). This demonstrates that the SP58 protein of *B. plicatilis* was possibly useful as a biomarker of exposure to particular toxicants. In *B. calyciflorus*, the inhibited esterase activities were also measured in response to ten different toxicants (e.g., pentachlorophenol, phenol, dimethylphenol, naphthol, xylene, cadmium, copper, mercury, diazinon, chlorpyrifos) (Burbank and Snell 1994) and obtained significant results in a dose-dependent manner. The performance of in vivo enzyme tests is less labor intensive compared to traditional whole animal testing. Similarly, the inhibited esterase activities measured by the modulations of fluorescence were evaluated in *B. plicatilis* as a biomarker for monitoring the effect of six different toxicants, particularly the pesticide metabolism (e.g., PCP, TBT, copper, mercury, zinc, and $Ca(OCl)_2$) (Moffat and Snell 1995). This later study also supported that the inhibition of esterase is an attractive marker for assessing aquatic toxicity due to its speed, simplicity, sensitivity, and applicability to a broad range of aquatic species.

In other rotifer species *Lecane* sp. (*L. hamata*, *L. luna*, *L. quadridentata*), esterase inhibition also showed a great potential as a biomarker for assessing six metals (e.g., cadmium, chromium, copper, lead, mercury, chloride, and titanium) and four organics (e.g., benzene, ethyl acetate, toluene, and vinyl acetate) exposure with species and toxicant specificity (Pérez-Legaspi et al. 2002), indicating that the esterase activity can be an outstanding biomarker in *L. luna* and *L. quadridentata* in response to metal exposure.

Recently, the mechanistic understanding of the antioxidant system was applied to uncover how rotifers respond to toxicants as physiological biomarkers. For example, in the rotifer *B. koreanus*, nonenzymatic biomarkers, glutathione (GSH) and reactive oxygen species (ROS) levels, were effectively applied to measure the state of oxidative stress in response to toxicant exposure including copper, hydrogen peroxide (H_2O_2), and radiation (Table 10.4). Particularly, ROS levels indicated cellular

Table 10.4 The enzyme activities measured in rotifer from recent published studies on the euryhaline rotifer *Brachionus sp.* for ecotoxicology and environmental monitoring studies

Species	Chemicals or Environmental stress	Endpoint	Remarks	Reference
B. koreanus	UV radiation (100 μW/cm^2, 0, 1, 2 and 4kJ/m^2)	ROS	Bell shape increasing at 2kJ/m^2 and decrease to control levels at 4 kJ/m^2	Kim et al. (2011)
		GSH		
		GPx		
		GST		
		GR		
	Gamma radiation (0, 50, 100, 150 and 200 Gy)	ROS	Highest at 150 Gy	Han et al. (2014)
		GST	Highest at 100 Gy	
	Triphenyltin	ROS	Increase at 5 μg/L	Yi et al. (2016)
	Copper (0.12 and 0.24 mg/L)	ROS	Dose dependently increased	Han et al. (2013)
		GSH		
		GPx		
		GR		
		GST		
	ATP /ATN/ CBZ/ OTC/ SMX/ TMP[a]	AChE	Dose dependently decreased	Rhee et al. (2013a)
	B[α]P (10, 100 μg/L)	GST	Dose and time dependently increased	Kim et al. (2013)
		SULT		
		GSH		
		GPx		
		GR		
		SOD		
	H_2O_2	GR	Dose dependently increased	Rhee et al. (2011)
		GPx		
		GST	Dose dependently decreased	
		GSH		
B. plicatilis	PCP/TBT/Copper/Mercury/ Zinc/Ca(OCl)$_2$	Esterase	Dose dependently decreased	Moffat and Snell (1995)

[a] *ATP* acetaminophen, *ATN* atenolol, *CBZ* carbamazepine, *OTC* oxytetracycline, *SMX* sulfamethoxazole, *TMP* trimethoprim, *PCP* pentachlorophenate, *TBT* tributyltin

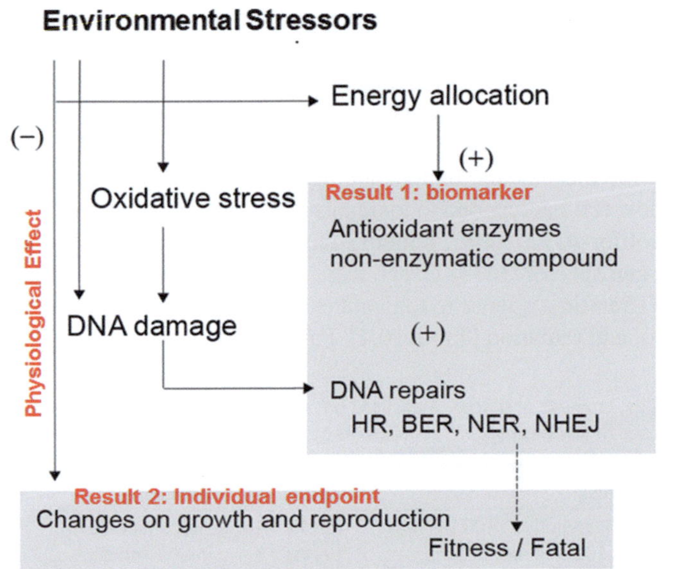

Fig. 10.3 Schematic diagram of the modulated patterns of physiological effects and energy allocation for detoxification and repair mechanisms in response to environmental stressors in marine organism (Adopted from Won et al. 2015; *HR* homologous recombination, *BER* base excision repair, *NER* nucleotide excision repair, *NHEJ* nonhomologous end joining)

oxidative levels induced by unbalanced conditions between the generation and removal mechanisms of ROS in damaged cells in response to environmental stress (Kim et al. 2011; Rhee et al. 2011; Han et al. 2013).

Considering oxidative stress and the response of its defensome, the response of glutathione *S*-transferase (GST), glutathione reductase (GR), and superoxide dismutase (SOD) was evaluated in rotifers in response to environmental stressors and toxicants (e.g., TPT, H_2O_2, copper, B[*a*]P, UV radiation) (Table 10.4) (Kim et al. 2011, 2013; Rhee et al. 2011; Han et al. 2013; Yi et al. 2016). Also, biochemical analyses can support biological alterations (e.g., reproductive impairment associated with population dynamics and reduced survival rate) of rotifers observed at individual levels under lab-based stress conditions. Thus, environmental stressors can cause several physiological alterations including unbalanced redox conditions and DNA damages in marine organisms. Then, the damage repair, regulation, and antioxidant mechanisms activated by energy allocation (e.g., energy for growth and reproduction) from normal conditions can regulate the fitness and/or other fates of organisms (Fig. 10.3). For example, copper and B[*a*]P-induced oxidative stress measured by enzymatic activities of SOD, glutathione peroxidase (GPx), GST, and GR were directly associated with retardation of growth and reproduction in *B. koreanus* (Han et al. 2013; Kim et al. 2013).

10.4 Molecular Endpoints of Rotifers for Ecotoxicological Studies as Biomarkers

10.4.1 Modulated Transcriptomes as Biomarkers

Since the early 2000s, genomic information in diverse non-model organisms have been accumulated with the development of NGS technology. Particularly, NGS and advanced bioinformatic tools made it easy to mine useful genomic information with less money for a short time, allowing that non-model species can be applied for marine environmental researches.

Using these genomic data, molecular and ecotoxicological studies from non-model organisms were easily carried out. Thus, whole genome- and RNA-seq-based studies have been performed in non-model aquatic invertebrates, rendering comparative analysis with vertebrates in terms of the mode and mechanisms of action in ecotoxicology, ecophysiology, and environmental monitoring. The genetic homogeneity of rotifers are of great strength in ecotoxicology, ecophysiology, and mechanistic studies using genomic data, as there is less noise to obtain the robust expression data in response to environmental pollutants.

To date, ecotoxicological studies with rotifers have continually been conducted on the base of rotifer biology (Snell and Janssen 1995; Zhang et al. 2013). In a recent review of rotifers, several advantages of rotifers in ecotoxicology were already discussed (Dahms et al. 2011). However, only a few studies were available using rotifer genomic information. Until 2014, a total of 25 Science Citation Index (SCI) papers have been published, focusing on ecotoxicology using genomic information of rotifers.

The quantitative analysis of stress-associated gene expression can provide perspectives to evaluate molecular responses in environmental risk assessment (ERA). For example, changes of mRNA expression patterns of heat shock protein (Hsp) and ubiquitin-conjugating enzyme were examined for use as an index for biomarkers for monitoring population growth of the rotifer *B. plicatilis* (Kaneko et al. 2002). Also recently, adverse outcome pathways (AOPs) showed great promise as a useful tool for predicting adverse outcomes observed in organisms (e.g., cancer, embryo malformation, death) in parallel with molecular and cellular endpoints in evaluating ERA based on adverse outcome and molecular biomarkers (Lee et al. 2015b). In particular, rotifers are good models to develop an AOP, linking the molecular response to effects on the individual and population levels in ERA, as genomic DNA and RNA-seq databases of rotifers provide a wealth of information of diverse molecular initiating events (MIEs) in response to environmental pollutant exposure.

10.4.2 Examples of Molecular Biomarkers in the Rotifer Brachionus koreanus

Species of the rotifer genus *Brachionus* sp. have been used as a model species in ecotoxicological studies with several favorable characteristics (Cochrane et al. 1991; Gama-Flores et al. 2004; Kim et al. 2011; Han et al. 2014). Among them, transcriptomic studies using *B. koreanus* were conducted to show the potency of rotifers as a model taxon that allows covering diverse fields including ecotoxicology, toxicogenomics, and ecophysiology (Lee et al. 2011, 2015a) to check the modulation of several detoxifying, metabolic, and DNA repair-associated genes as molecular biomarkers for particular environmental stressors (Table 10.5). Briefly, mRNA expression patterns of 25 cytochrome P450 (*CYP*) genes were examined as indices for molecular biomarkers monitoring B[*a*]P (Kim et al. 2013). They could provide indications that *CYP* genes would play an important role in metabolizing organic pollutants such as B[*a*]P a to find mechanisms of toxicity. Also, the modulated expressions of phase II mechanisms (e.g., *GSTs*) and antioxidant enzymes (e.g., *GPx*, *CuZnSOD*, *MnSOD*, and *catalase*) with *hsp* and DNA repair genes were used to measure molecular and cellular damages in response to environmental stressors (e.g., B[*a*]P, copper, UV, and gamma radiation) (Kim et al. 2011, 2013; Han et al. 2013, 2014).

10.4.2.1 Cytochrome P450 Genes

Whole 25 *CYP* genes from rotifers were firstly identified from the rotifer *B. koreanus* (Kim et al. 2013). CYP enzymes have prominent roles in the metabolism of xenobiotics and the biotransformation and catalysis of endogenous compounds such as steroids, fatty acids, and diverse hormones (Snyder 1998). Thus, responses of *CYP* genes or proteins in response to environmental pollutants are considered as biomarkers of pollution in aquatic ecosystem (Bucheli and Fent 1995; Goksøyr 1995). Among rotifers, only a little information is available on *CYP* genes but the different function of each CYP isoform is widely studied in vertebrates. This holds particularly for the mode of action on the toxic effect of CYP-associated compounds and, particularly, the aryl hydrocarbon receptor (AhR)-mediated induction of the CYP1 family in the biotransformation of specific pollutants such as β-naphthoflavone, B[*a*]P, and other PAHs. For example, four different CYP1 families (CYP1A, CYP1B, CYP1C, and CYP1D) are considered as the most important chemical defense systems with AhR. However, in invertebrates, the CYP1 family is not existing. Thus, the function of each CYP family in invertebrates may provide new information on how invertebrates regulate and metabolize toxicants in their respective systems to detoxify.

In the rotifer *B. koreanus*, the whole set of 25 *CYP* genes were firstly identified, and their modulated expressions were measured in response to B[*a*]P exposure

Table 10.5 Molecular biomarker genes and their functions in recent published studies on the euryhaline rotifer *Brachionus koreanus* for ecotoxicology and environmental monitoring studies

Function		Gene	Organic pollutant		Physical stress	Oxidants (or trace metals)			Radiation		Pharmaceuticals
			B(a)P	TPT	Heat	H_2O_2	Cu	Cd	Gamma ray	UV	ATP/ATN/CBZ/OTC/SMX/TMP[a]
Development		*DMRT*	X							X	
Neural system		*AChE*									X
Chaperoning		*Hsps*		X	X	X			X	X	
							X	X			
Detoxification	Phase I	*CYPs*	X								
				X							
	Phase II	*GSTs*	X	X			X	X	X		
	Phase III	*Pgp*			X						X
		SULT	X								
	Antioxidant	*SOD, CAT*	X								
DNA repair	HR[a]	*RAD*							X		
	NHEJ[a]	*Ku, DNA-pk, DNA ligase*							X		
										X	
	NER[a]	*RPA, XPV*							X		
										X	
	SSB[a]	*XRCC*							X		
	MMR[a]	*MSH, MLH, PCNA*							X		

[a]*HR* homologous recombination, *NHEJ* nonhomologous end joining, *NER* nucleotide excision repair, *SSB* single-strand DNA binding, *MMR* mismatch repair, *ATP* acetaminophen, *ATN* atenolol, *CBZ* carbamazepine, *OTC* oxytetracycline, *SMX* sulfamethoxazole, *TMP* trimethoprim

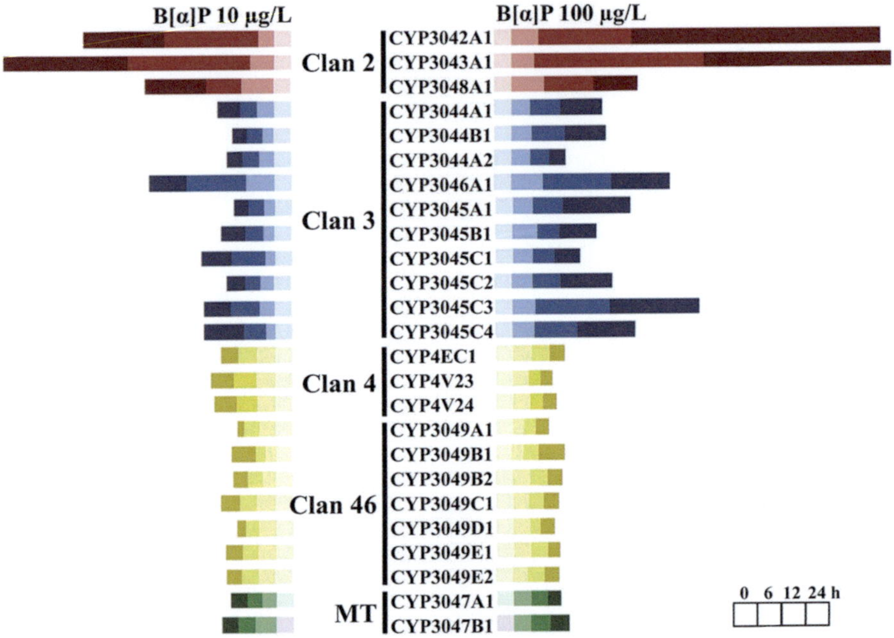

Fig. 10.4 The transcript profiles of *B. koreanus* whole *CYP* genes in response to the exposure of two concentrations of B[*a*]P exposure for 24 h (Adopted from Kim et al. 2013)

(Fig. 10.4). Also, the mode of action of B[*a*]P has been reported with full sequences of phase I biotransformation enzymes with respect to the molecular defense metabolisms (Kim et al. 2013). In 25 rotifer *CYP* genes, they were separated with five distinct clans (clan 2, clan 3, clan 4, clan 46, and a mitochondrial clan). Of them, three *CYP* genes (e.g., *CYP3042A1*, *CYP3043A1*, and *CYP3048A1*) belong to clan 2 that was composed of the vertebrate CYP1 family, insect CYP307, and *Daphnia* CYP364 family. These rotifer three *CYP* genes are likely to play similar functions as the vertebrate CYP1 in the B[*a*]P metabolism. Also ten rotifer *CYP* genes in clan 3 were clustered with vertebrate CYP3 and *Daphnia* CYP361 families that are putatively involved in the detoxification and the thromboxane A2 biosynthesis, respectively (Baldwin et al. 2009).

10.4.2.2 Phase II Detoxification and Antioxidant Defense Mechanism Genes

Phase II biotransformation including UDP-glucuronosyltransferases, sulfotransferases (SULTs), N-acetyltransferases, GSTs, various methyltransferases, and catechol O-methyl transferase plays an important role in conjugation, following phase I mechanisms such as oxidative, reductive, and hydrolytic reactions by *CYP* genes. In the rotifer *B. koreanus*, a series of detoxification of B[*a*]P was firstly reported on phase II biotransformation mechanism at the transcriptional level and enzymatic activities of GST and SULT (Kim et al. 2013). Significantly induced mRNA levels and enzymatic activities in GSTs and SULT in response to B[*a*]P exposure (10 and 100 μg/L) indicated that antioxidant activity can be induced by B[*a*]P exposure, suggesting that B[*a*]P was metabolized by conjugating GSH with oxidized functional groups through phase I and phase II detoxification mechanisms in *B. koreanus*.

The modulation of mRNA expression with antioxidant enzymes turned out to be useful in examining how organisms respond to radiation-induced ROS and also to show how physiological alterations of rotifers correlated with the individual and population levels (e.g., growth retardation and reduced survival rate) in response to gamma and UV radiation (Kim et al. 2011; Han et al. 2014). Regarding oxidative stress, all molecular biomarkers (*GST*, *GPx*, *MnSOD*, *CuZnSOD*, *CAT*) associated with oxidative stress were sensitively responding to ROS levels in response to B[*a*]P in the rotifer *B. koreanus* (Kim et al. 2013). Dramatic increase of *GST-omega*, *GST-sigma*, and *GST-zeta* genes in copper-exposed *B. koreanus* showed that GSTs were induced by Cu exposure as one of the enzymatic defense mechanisms, particularly in the early stage of oxidative stress response (Han et al. 2013). Thus, the phase II detoxification and antioxidant defense mechanism genes have a potential as biomarkers for a more sensitive stress response at the initial stage with a high correlation between mRNA and their related enzymatic activities.

10.4.2.3 Heat Shock Protein (Hsp) Genes

Heat shock proteins (Hsps) play a role in protein homeostasis by regulating the protein folding (Nollen and Morimoto 2002; Imai et al. 2003). Hsps have been used as markers for cellular defense mechanisms following the discovery of denatured protein chaperoning and the degradation of proteins by stress-induced damage (Feder and Hofmann 1999). Rotifer *hsp* genes were identified as shown in Table 10.6. Effective RNAi-mediated suppression of 4 *hsp* genes showed that these are essential for survival and adaptation to thermal stress in rotifers (Smith et al. 2012). In *B. koreanus*, 12 different *hsp* genes were identified by EST and NGS techniques (Lee et al. 2011; Kim et al. 2011). Their expressional modulations were examined in response to Cu, Cd, B[*a*]P, TPT, and radiation (Kim et al. 2011; Jung and Lee 2012; Han et al. 2014; Yi et al. 2016). Of them, *hsp70* expression was consistently upregulated in response to chemical exposures. However, dramatic increases of *hsp* 90α2 and *hsp 40* genes were shown as defense mechanisms for TPT and for gamma irradiation-induced stress, suggesting that each *hsp* gene has a distinctive role in chaperoning proteins (Yi et al. 2016). Also, antioxidant function of *Hsp20* genes was shown for enhancing cell survival in H_2O_2-exposed *Brachionus* sp. through a disk assay (Rhee et al. 2011). Briefly, *Brachionus Hsp20* expressed in *E.coli* showed higher viability than that of only vector-containing *E.coli* after H_2O_2 exposure, indicating that rotifer *Hsp20* genes play a protective role in response to oxidative stress (Rhee et al. 2011). In conclusion, elevated expression of *hsp* genes provides a crucial function in the protection from oxidative stress and/or DNA repair processes in response to chemical exposure in rotifers as it was shown earlier in vertebrates.

Table 10.6 Heat shock protein (*hsp*) genes from rotifers

Species	Name	GenBank accession no.	References
B. koreanus	*Hsp20*	GU461594	Lee et al. (2011)
	Hsp27	GU574481	Kim et al. (2011)
	Hsp70	GU574486	
	Hsc70	GU574487	
	Hsp90α1	GU574488	
	Hsp90β	GU574490	
	Hsp10	GU574479	
	Hsp21	GU574480	
	Hsp30	GU574482	
	Hsp40	GU574483	
	Hsp40h	GU574484	
	Hsp60	GU574485	
	Hsp90α2	GU574489	
B. plicatilis	*Hsp70*	AB076052	Kaneko et al. (2002)
B. manjavacas	*hsp 40*	HQ901983	Smith et al. (2012)
	hsp 60	HQ901985	
	hsp 70–3	HQ901984	
	hsp 90	HQ901986	
Plationus patulus	*HSP 60*		Rios-Arana et al. (2005)

10.4.2.4 DNA Repair Genes

Ionizing radiation (IR) is important to enhance DNA damage (e.g., single- and double-strand DNA breaks, basic sites, and alterations of DNA bases) (Ward and Kuo 1976; Rhee et al. 2013b). In gamma-irradiated *B. koreanus*, several DNA repair-associated genes were significantly increased (Fig. 10.5) with repercussions (e.g., growth retardation and impairment in reproduction) at the individual level.

Also, their life spans were significantly reduced (Han et al. 2014). Thus, a causal correlation of gamma radiation was demonstrated with individual parameters (e.g., survival rate, life span, fecundity, growth retardation) and several molecular biomarkers associated with DNA damage and antioxidant defense mechanisms, suggesting that *B. koreanus* recovers oxidative stress-induced cellular and DNA damage caused by gamma radiation through subsequent defense mechanisms using antioxidants (*GST-sigma*, *GST-omega*, and *GPx*), chaperoning processes (heat shock protein genes, *hsp 40*, *hsp 70*, *hsp 90a1*), and DNA repair pathways. In *B. koreanus*, UV-B radiation would affect up- or downregulation of DNA replication and repair process (e.g., *RPA*, *DNA-PK*, *Ku70*, and *Ku80*) with an alteration of chaperoning genes, leading to growth retardation (Kim et al. 2011). In particular, significant and fast increase of *DNA-PK* and *Ku70/80* genes indicated that these genes are involved in repairing processes in response to a low dose of UV-B exposure (2 kJ/m^2).

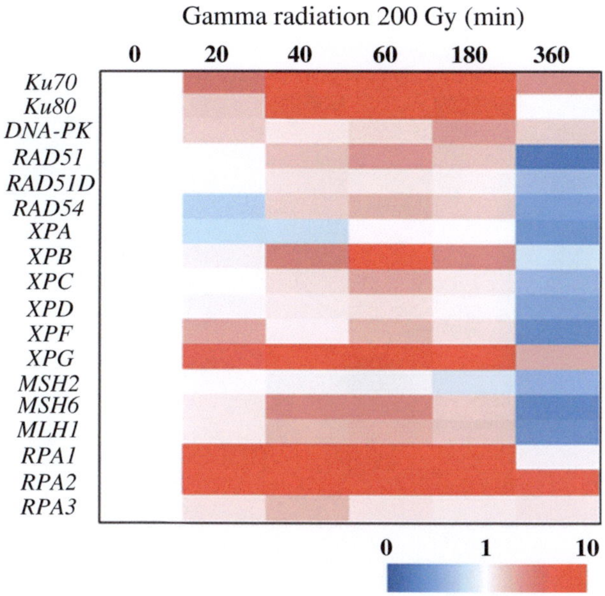

Fig. 10.5 Effects of gamma irradiation (200 Gy, 2 Gy/min) on relative mRNA expressions of DNA repair-related genes in the rotifer *B. koreanus*. Modulations of mRNA expressions were measured at postirradiation with time courses (0, 20, 40, 60, 180, and 360 min) (Adopted from Han et al. 2014)

10.5 Forthcoming Tools Provided by Rotifers for Ecotoxicology

In general, protocols of several endpoints on mortality, growth retardation, reproduction, behavior, and cellular biomarkers have been published and are being used by several studies in response to toxicant exposure. Of them, recent developments in molecular techniques for rotifer have expanded our mechanistic understanding of chemical toxicity at the molecular level, possibly linking to population levels for ecological relevance. In particular, whole genome sequencing is considered as an emerging technique for mining enormous genomic data from rotifers, which provide a better understanding of the evolution, physiology, and ecotoxicology across animal phyla and their evolution.

In situ hybridization fluorescence analysis was applied in rotifers (Boell and Bucher 2008; Smith et al. 2010). A whole-mount of a tiny invertebrate such as a rotifer in situ hybridization provides several merits to identify a location of gene expressions. Particularly, small and transparent organisms such as rotifer can be easily applied for such functional analysis of genes. In *B. plicatilis*, germ cell marker genes (e.g., *vasa* and *nanos*) were identified spatiotemporally (Smith et al. 2010). Also, *B. koreanus vasa* gene expression was identified through the development of *B. koreanus* eggs (Fig. 10.6), suggesting that *B. koreanus vasa* genes are associated

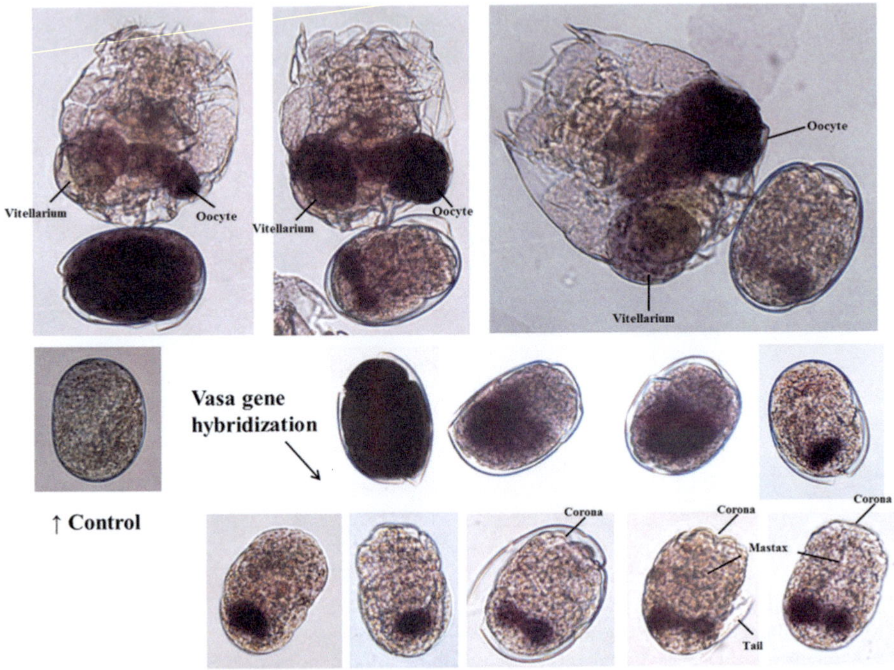

Fig. 10.6 (**a**) Expression of *vasa* genes in adult oocytes. (**b**) Expression of *vasa* genes in the *Brachionus koreanus* egg over developmental stages of *B. koreanus* embryos (**a–j**) (Kim et al. unpublished data)

with germ cell development and can directly be applied for the examination of reproductive impairment in response to environmental stressors.

For examining activation of signal transduction pathways in response to cellular damage, Western blot analysis was also introduced in rotifers. For example, in multi-walled carbon nanotube (MWCNTs)-exposed *B. koreanus*, activation of mitogen-activated protein kinase signaling pathways, phosphorylating extracellular signal-regulated kinases (ERK), JNK, and p38, was examined using Western blot analysis (Lee et al. 2016). In this study, the blots developed with a peroxidase-conjugated mammalian antibody (ERK from mouse and all other genes from rabbit) indicated that the rotifers have conserved signal pathway genes that can bind with mammalian antibodies and respond in response to cellular damage (Fig. 10.7).

Using RNA interference allows elucidating the physiological function of genes through the knockdown of target genes (Fig. 10.8). It also enables to maximize the utilization for functional studies. In *B. manjavacas*, knockdown of progesterone hormone receptors using RNA interference indicated that the function of progesterone was conserved with vertebrates (Stout et al. 2010).

Since 1960s, an enormous amount of rotifer studies has filled the knowledge gap of molecular biology/biochemistry and ecology in response to environmental stress. Recent development in techniques for sequestering new information using rotifers

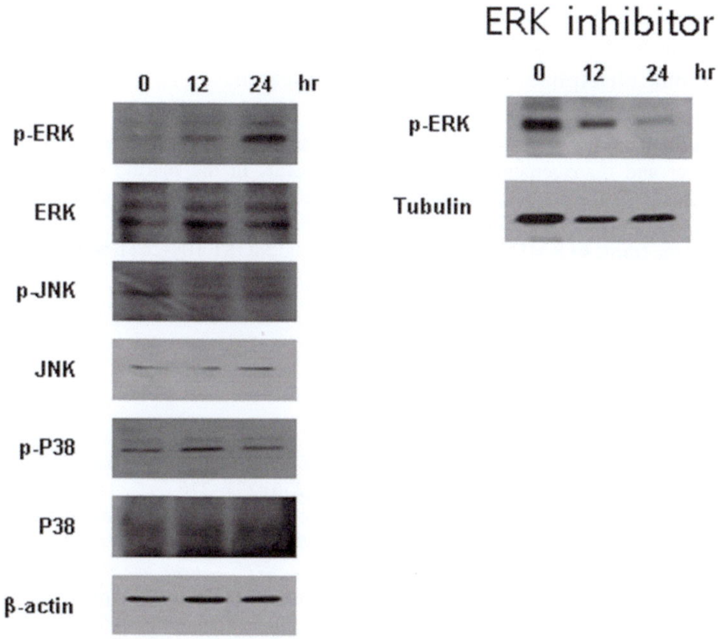

Fig. 10.7 Time-dependent mitogen-activated protein kinase (MAPK) protein expression levels of MWCNT-exposed *B. koreanus* over a period of 24 h (Adopted from Lee et al. 2016)

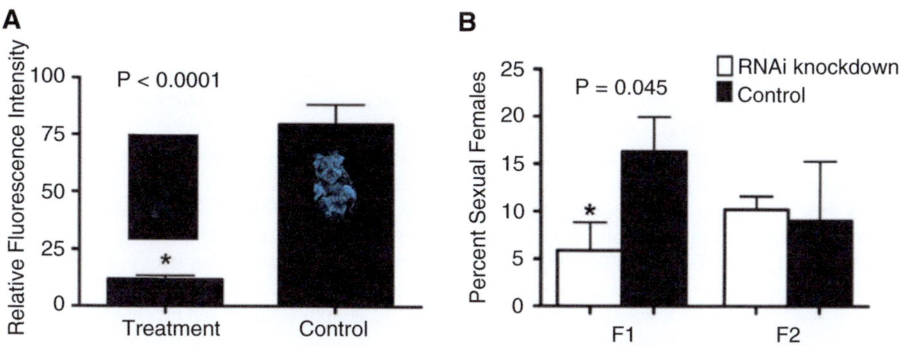

Fig. 10.8 RNAi experiments with female rotifers transfected with dsRNA from the rotifer progesterone receptor gene (treatment), dsRNA from the rotifer elongation factor gene (control for **a**), or with PBS (control for **b**). (**a**) Relative fluorescence intensity of female rotifers incubated with a progesterone probe ($n = 6$, two sample, one-tailed paired student's t test, $P < 0.0001$; *error bars* denote SE). (**b**) Percent sexual females in first-generation rotifer daughters (F1) and second-generation rotifer daughters (F2) ($n = 4$, two sample, one-tailed paired student's t test, $P = 0.045$ for F1; $n = 4$, two sample, one-tailed paired student's t test, $P = 0.434$ for F2; *error bars* denote SE) (Adopted from Stout et al. 2010)

allows that rotifers can be a part of a battery of ecotoxicological tests. Most of all, future work in ecotoxicological studies with rotifer genome information will benefit integrative multidisciplinary and sustainable research.

Acknowledgments This work was supported by a grant (20140342, Development of techniques for assessment and management of hazardous chemicals in the marine environment) of the Ministry of Oceans and Fisheries, Korea, funded to Jae-Seong Lee. H-U D acknowledges the support of the Asia-Pacific Ocean Research Center of the Department of Oceanography (No. 76211194) in the frame of KMU/NSYSU cooperation and MOST 104-2621-M-037-001 funded to T.H. SHIH.

References

Arndt H (1993) Rotifers as predators on components of the microbial web (bacteria, heterotrophic flagellates, ciliates) – a review. Hydrobiologia 255(256):231–246

Arnold WR, Diamond RL, Smith DS (2010) The effects of salinity, pH, and dissolved organic matter on acute copper toxicity to the rotifer, *Brachionus plicatilis* ("L" strain). Arch Environ Contam Toxicol 59:225–234

Arnold WR, Diamond RL, Smith DS (2011) Acute and chronic toxicity of copper to the euryhaline rotifer, *Brachionus plicatilis* ("L" strain). Arch Environ Contam Toxicol 60:250–260

Austad SN (2009) Is there a role for new invertebrate models for aging research? J Gerontol A 64:192–194

Baldwin WS, Marko PB, Nelson DR (2009) The cytochrome P450 (CYP) gene superfamily in *Daphnia pulex*. BMC Genomics 10:169

Boell L, Bucher G (2008) Whole-mount in situ hybridization in the rotifer *Brachionus plicatilis* representing a basal branch of lophotrochozoans. Dev Genes Evol 218:445–451

Bucheli TD, Fent K (1995) Induction of cytochrome P450 as a biomarker for environmental contamination in aquatic ecosystems. Crit Rev Environ Sci Technol 25:201–268

Burbank SE, Snell TW (1994) Rapid toxicity assessment using esterase biomarkers in *Brachionus calyciflorus* (Rotifera). Environ Toxicol 9:171–178

Capuzzo JM (1979) The effect of halogen toxicants on survival, feeding, and egg production of the rotifer *Brachionus plicatilis*. Estuar Coast Mar Sci 8:307–316

Clark MS, Denekamp NY, Thorne MAS, Reinhardt R, Drungowski M (2012) Long-term survival of hydrated resting eggs from *Brachionus plicatilis*. PLoS One 7:e29365

Cochrane BJ, Irby RB, Snell TW (1991) Effects of copper and tributyltin on stress protein abundance in the rotifer *Brachionus plicatilis*. Comp Biochem Physiol 98C:385–390

Dahms H-U, Hagiwara A, Lee J-S (2011) Ecotoxicology, ecophysiology, and mechanistic studies with rotifers. Aquat Toxicol 101:1–12

Denekamp NY, Thorne MAS, Clark MS, Kube M, Reinhardt R, Lubzens E (2009) Discovering genes associated with dormancy in the monogonont rotifer *Brachionus plicatilis*. BMC Genomics 10:108

Feder ME, Hofmann GE (1999) Heat-shock proteins, molecular chaperones, and the stress response: evolutionary and ecological physiology. Annu Rev Physiol 61:243–282

Ferrando MD, Janssen CR, Andreu E, Persoone G (1993) Ecotoxicological studies with the freshwater rotifer *Brachionus calyciflorus*. III. The effects of chemicals on the feeding behavior. Ecotoxicol Environ Saf 26:1–9

Gama-Flores JL, Sarma SSS, Nandini S (2004) Acute and chronic toxicity of the pesticide methyl parathion to the rotifer *Brachionus angularis* (Rotifera) at different algal (*Chlorella vulgaris*) food density. Aquat Ecol 38:27–36

Gannon JE, Stemberger RS (1978) Zooplankton (especially crustaceans and rotifers) as indicators of water quality. Trans Am Microsc Soc 97:16–35

Goksøyr A (1995) Use of cytochrome P450 1A (CYP1A) in fish as a biomarker of aquatic pollution. Arch Toxicol Suppl 17:80–95

Gomez A, Serra M, Carvalho GR, Lunt DH (2002) Speciation in ancient cryptic species complexes: evidence from the molecular phylogeny of *Brachionus plicatilis* (Rotifera). Evolution 56:1431–1444

Gutkowska A, Paturej E, Kowalska E (2013) Rotifer trophic state indices as ecosystem indicators in brackish coastal waters. Oceanologia 55:887–899

Guzzella L, Gronda A, Colombo L (1997) Acute toxicity of organophosphorus insecticides to marine invertebrates. Bull Environ Contam Toxicol 59:313–320

Hagiwara A (1994) Practical use of rotifer cysts. Israel J Aquacult 46:13–21

Hagiwara A, Kotani T, Snell TW, Assava-Aree M, Hirayama K (1995) Morphology, reproduction, genetics, and mating behavior of a small, tropical marine *Brachionus* strain (Rotifera). J Exp Mar Biol Ecol 194:25–37

Halbach U (1984) Population dynamics of rotifers and its consequences for ecotoxicology. Hydrobiologia 109:79–96

Halbach U, Weibert M, Wissel C, Klaus J, Beuter K, Delion M (1981) Population dynamics of rotifers as bioassay tools for toxic effects of organic pollutants. Verh Int Ver Limnol 21:1141–1146

Han J, Won E-J, Hwang D-S, Rhee J-S, Kim I-C, Lee J-S (2013) Effect of copper exposure on GST activity and on the expression of four *GST*s under oxidative stress condition in the monogonont rotifer, *Brachionus koreanus*. Comp Biochem Physiol C 158:91–100

Han J, Won E-J, Kim I-C, Yim J-H, Lee S-J, Lee J-S (2014) Sublethal gamma irradiation affects reproductive impairment and elevates antioxidant enzyme and DNA repair activities in the monogonont rotifer *Brachionus koreanus*. Aquat Toxicol 155:101–109

Imai J, Yashiroda H, Maruya M, Yahara I, Tanaka K (2003) Proteasomes and molecular chaperones. Cell Cycle 2:585–589

Isidori M, Lavorgna M, Nardelli A, Pascarella L, Parrella A (2005) Toxic and genotoxic evaluation of six antibiotics on non-target organisms. Sci Total Environ 346:87–98

Juchelka CM, Snell TW (1994) Rapid toxicity assessment using rotifer ingestion rate. Arch Environ Contam Toxicol 26:549–554

Jung M-Y, Lee Y-M (2012) Expression profiles of heat shock protein gene families in the monogonont rotifer *Brachionus koreanus* – exposed to copper and cadmium. Toxicol Environ Health Sci 4:235–242

Kaneko G, Kinoshita S, Yoshinaga T, Tsukamoto K, Watabe S (2002) Changes in expression patterns of stress protein genes during population growth of the rotifer *Brachionus plicatilis*. Fish Sci 68:1317–1323

Kim H, Hagiwara A (2009) Aging and viability of resting eggs in the rotifer *Brachionus plicatilis* Müller. In: Rotifera XII, Berlin, 16–21 August 2009, p. 96

Kim R-O, Rhee J-S, Won E-J, Lee K-W, Kang C-M, Lee Y-M, Lee J-S (2011) Ultraviolet B retards growth, induces oxidative stress, and modulates DNA repair-related gene and heat shock protein gene expression in the monogonont rotifer, *Brachionus* sp. Aquat Toxicol 101:529–539

Kim R-O, Kim B-M, Jeong C-B, Nelson DR, Lee J-S, Rhee J-S (2013) Expression pattern of entire cytochrome P450 genes and response of defensomes in the benzo[*a*]pyrene-exposed monogonont rotifer *Brachionus koreanus*. Environ Sci Technol 47:13804–13812

Kim B-M, Jeong C-B, Kim I-C, Yim J-H, Lee Y-S, Rhee J-S, Lee J-S (2014a) Identification of three doublesex genes in the monogonont rotifer *Brachionus koreanus* and their transcriptional responses to environmental stressor-triggered population growth retardation. Comp Biochem Physiol 174B:36–44

Kim H-J, Sawada C, Hagiwara A (2014b) Behavior and reproduction of the rotifer *Brachionus plicatilis* species complex under different light wavelengths and intensities. Int Rev Hydrobiol 99:151–156

Krisko A, Leroy M, Radman M, Meselson M (2012) Extreme anti-oxidant protection against ionizing radiation in bdelloid rotifers. Proc Natl Acad Sci U S A 109:2354–2357

Lee J-S, Rhee J-S, Kim R-O, Choi B-S, Park HG, Park GS, Hagiwara A, Choi IY (2011) Genomic DNA information (680 Mb) and molecular insights to the monogonont rotifer *Brachionus ibericus*. Hydrobiologia 662:65–75

Lee B-Y, Kim H-S, Hwang D-S, Won E-J, Choi B-S, Choi I-Y, Park HG, Rhee J-S, Lee J-S (2015a) Whole transcriptome analysis of the monogonont rotifer *Brachionus koreanus* provides molecular resources for developing biomarkers of carbohydrate metabolism. Comp Biochem Physiol 14D:33–41

Lee JW, Won E-J, Raisuddin S, Lee J-S (2015b) Significance of adverse outcome pathways in biomarker-based environmental risk assessment in aquatic organisms. J Environ Sci 35:115–127

Lee JW, Kang H-M, Won E-J, Hwang D-S, Kim D-H, Lee S-J, Lee J-S (2016) Multi-walled carbon nanotubes (MWCNTs) lead to growth retardation, antioxidant depletion, and activation of the ERK signaling pathway but decrease copper bioavailability in the monogonont rotifer *Brachionus koreanus*. Aquat Toxicol 172:67–79

Marcial HS, Hagiwara A, Snell TW (2005) Effect of some pesticides on reproduction of rotifer *Brachionus plicatilis* Müller. Hydrobiologia 546:569–575

Moffat BD, Snell TW (1995) Rapid toxicity assessment using in vivo enzyme test for *Brachionus plicatilis* (Rotifera). Ecotoxicol Environ Saf 30:47–53

Nollen EA, Morimoto RI (2002) Chaperoning signaling pathways: molecular chaperones as stress-sensing 'heat shock' proteins. J Cell Sci 115:2809–2816

Ortells R, Snell TW, Gomez A, Serra M (2000) Patterns of genetic differentiation in resting egg banks of a rotifer species complex in Spain. Arch Hydrobiol 149:529–551

Ortells R, Gomez A, Serra M (2003) Coexistence of cryptic rotifer species: ecological and genetic characterisation of *Brachionus plicatilis*. Freshw Biol 48:2194–2202

Pérez-Legaspi IA, Rico-Martínez R, Pineda-Rosas A (2002) Toxicity testing using esterase inhibition as a biomarker in three species of the genus *Lecane* (Rotifera). Environ Toxicol Chem 21:776–782

Persoone G, Blaise C, Snell T, Janssen C, van Steertegem M (1993) Cyst-based toxicity tests: 2. – report on an international intercalibration exercise with three cost-effective toxkits. Zeit Angew Zool 78:17–36

Preston BL, Snell TW (2001) Full life-cycle toxicity assessment using rotifer resting egg production: implications for ecological risk assessment. Environ Pollut 114:399–406

Preston BL, Snell TW, Fields DM, Weissburg MJ (2001) The effects of fluid motion on toxicant sensitivity of the rotifer *Brachionus calyciflorus*. Aquat Toxicol 52:117–131

Preston BL, Cecchine G, Snell TW (1999a) Effects of pentachlorophenol on predator avoidance behavior of the rotifer *Brachionus calyciflorus*. Aquat Toxicol 44:201–212

Preston BL, Snell TW, Kneisel R (1999b) UV-B exposure increase acute toxicity of pentachlorophenol and mercury to the rotifer *Brachionus calyciflorus*. Environ Pollut 106:23–31

Pynnönen K (1996) Heavy metal-induced changes in the feeding and burrowing behaviour of a Baltic isopod, *Saduria (Mesidotea) entomon* L. Mar Environ Res 41:145–156

Radix P, Severin G, Schramm K-W, Kettrup A (2002) Reproduction disturbances of *Brachionus calyciflorus* (rotifer) for the screening of environmental endocrine disrupters. Chemosphere 47:1097–1101

Rego D, Redondo LM, Geraldes V, Costa L, Navalhoa J, Pereira MT (2015) Control of predators in industrial scale microalgae cultures with pulsed electric fields. Bioelectrochemistry 103:60–64

Rhee J-S, Kim R-O, Choi HG, Lee J, Lee Y-M, Lee J-S (2011) Molecular and biochemical modulation of heat shock protein 20 (*Hsp20*) gene by temperature stress and hydrogen peroxide (H_2O_2) in the monogonont rotifer, *Brachionus* sp. Comp Biochem Physiol 154C:19–27

Rhee J-S, Kim B-M, Jeong C-B, Park HG, Leung KM, Lee Y-M, Lee J-S (2013a) Effect of pharmaceuticals exposure on acetylcholinesterase (AchE) activity and on the expression of AchE gene in the monogonont rotifer, *Brachionus koreanus*. Comp Biochem Physiol 158C:216–224

Rhee J-S, Kim B-M, Kim R-O, Seo J-S, Kim I-C, Lee Y-M, Lee J-S (2013b) Co-expression of antioxidant enzymes with expression of *P53*, DNA repair, and heat shock protein genes in the gamma ray-irradiated hermaphroditic fish *Kryptolebias marmoratus* larvae. Aquat Toxicol 140(141):58–67

Rico-Martínez R, Snell TW, Shearer TL (2013) Synergistic toxicity of Macondo crude oil and dispersant Corexit 9500A® to the *Brachionus plicatilis* species complex (Rotifera). Environ Pollut 173:5–10

Rios-Arana JV, Gardea-Tprresdey JL, Webb R, Walsh EJ (2005) Heat shock protein 60 (HSP60) response of *Plationus patulus* (Rotifera: Monogononta) to combined exposures of arsenic and heavy metals. Hydrobiologia 546:577–585

Roper DS, Hickey CW (1993) Behavioural responses of the marine bivalve *Macomona liliana* exposed to copper- and chlordane-dose sediments. Mar Biol 118:673–680

Sarma SSS, Ramírez Pérez T, Nandini S (2000) Comparison of the sensitivity of *Brachionus calyciflorus* and *Brachionus patulus* (Rotifera) to selected heavy metals under low and high food (*Chlorella vulgaris*) levels. Bull Environ Contam Toxicol 64:735–739

Schaefer ED, Pipes WO (1973) Temperature and the toxicity of chromate and arsenate to the rotifer *Philodina roseola*. Water Res 7:1781–1790

Segers H (2007) Global diversity of rotifers (Rotifera) in freshwater. In: Balian EV, Lévêque C, Segers H, Martens K (eds) Freshwater animal diversity assessment, Developments in Hydrobiology, vol 198. Springer, Dordrecht

Sha J, Wang Y, Chen H, Wang M, Wang H, Li H, Lei Q, Tang X (2015) Using population demographic parameters to assess impacts of two polybrominated diphenyl ethers (BDE-47, BDE-209) on the rotifer *Brachionus plicatilis*. Ecotoxicol Environ Saf 119:106–115

Smith JM, Cridge AG, Dearden PK (2010) Germ cell specification and ovary structure in the rotifer *Brachionus plicatilis*. EvoDevo 1:5

Smith HA, Burns AR, Shearer TL, Snell TW (2012) Three heat shock proteins are essential for rotifer thermotolerance. J Exp Mar Biol Ecol 413:1–6

Snell TW, Hicks DG (2011) Assessing toxicity of nanoparticles using *Brachionus manjavacas* (Rotifera). Environ Toxicol 26:146–152

Snell TW, Janssen CR (1995) Rotifers in ecotoxicology: a review. Hydrobiologia 313(314):231–247

Snell TW, Moffat BD (1992) A 2-d life cycle test with the rotifer *Brachionus calyciflorus*. Environ Toxicol Chem 11:1249–1257

Snell TW, Persoone G (1989a) Acute toxicity bioassay using rotifer. I. A test for brackish and marine environments with *Brachionus plicatilis*. Aquat Toxicol 14:65–80

Snell TW, Persoone G (1989b) Acute toxicity bioassay using rotifer. II. A freshwater test with *Brachionus rubens*. Aquat Toxicol 14:81–92

Snell TW, Moffat BD, Janssen C, Persoone G (1991a) Acute toxicity tests using rotifers. IV. The effects of cyst age, temperature and salinity on the sensitivity of *Brachionus calyciflorus*. Ecotoxicol Environ Saf 21:308–317

Snell TW, Moffat BD, Janssen C, Persoone G (1991b) Acute toxicity tests using rotifers. III. The effects of temperature, strain and exposure time on the sensitivity of *Brachionus plicatilis*. Environ Toxicol Water 6:63–75

Snell TW, Brogdon SE, Morgan MB (2003) Gene expression profiling in ecotoxicology. Ecotoxicology 12:475–483

Snell TW, Kubanek J, Carter W, Payne AB, Kim J, Hicks MK, Stelzer C-P (2006) A protein signal triggers sexual reproduction in *Brachionus plicatilis* (Rotifera). Mar Biol 149:763–773

Snyder MJ (1998) Cytochrome P450 enzymes belonging to the CYP4 family from marine invertebrates. Biochem Biophys Res Commun 249:187–190

Stegman JJ (1978) Influence of environmental contamination on cytochrome P-450 mixed function oxygenases in fish: implications for recovery in the wild harbor marsh. J Fish Board Can 35:658–674

Stelzer C-P, Snell TW (2003) Induction of sexual reproduction in *Brachionus plicatilis* (Monogononta, Rotifera) by a density-dependent chemical cue. Limnol Oceanogr 48:939–943

Stout EP, La Clair JJ, Snell TW, Shearer TL, Kubanek J (2010) Conservation of progesterone hormone function in invertebrate reproduction. Proc Natl Acad Sci U S A 107:11859–11864

Suatoni E, Vicario S, Rice S, Snell TW, Caccone A (2006) An analysis of species boundaries and biogeographic patterns in a cryptic species complex: the rotifer—*Brachionus plicatilis*. Mol Phylogenet Evol 41:86–98

Suga K, Welch DM, Tanaka Y, Sakakura Y, Hagiwara A (2007) Analysis of expressed sequence tags of the cyclically parthenogenetic rotifer *Brachionus plicatilis*. PLoS One 2:e671

Verschoor AM, Boonstra H, Meijer T (2005) Application of stable isotope tracers to studies of zooplankton feeding, using the rotifer *Brachionus calyciflorus* as an example. Dev Hydrobiol 181:535–549

Wang H, Tang X, Sha J, Chen H, Sun T, Wang Y (2015) The reproductive toxicity on the rotifer *Brachionus plicatilis* induced by BDE-47 and studies on the effective mechanism based on antioxidant defense system changes. Chemosphere 135:129–137

Ward JF, Kuo I (1976) Strand breaks, base release, and post irradiation changes in DNA γ-irradiated in dilute O_2–saturated aqueous solution. Radiat Res 66:485–498

Won E-J, Lee J-S (2014) Gamma radiation induces growth retardation, impaired egg production, and oxidative stress in the marine copepod *Paracyclopina nana*. Aquat Toxicol 150:17–26

Won E-J, Dahms H-U, Kumar KS, Shin K-H, Lee J-S (2015) An integrated view of gamma radiation effects on marine life: from molecules to ecosystems. Environ Sci Pollut Res 22:17443–17452

Yi AX, Han J, Lee J-S, Leung KM (2016) Toxicity of triphenyltin chloride to the rotifer *Brachionus koreanus* across different levels of biological organization. Environ Toxicol 31:13–23

Zhang L, Niu J, Li Y, Wang Y, Sun D (2013) Evaluation the sub-lethal toxicity of PFOS and PFOA using rotifer *Brachionus calyciflorus*. Environ Pollut 180:34–40

Index

A. Hagiwara, T. Yoshinaga (eds.), *Rotifers*, Fisheries Science Series,
DOI 10.1007/978-981-10-5635-2

MIX
Papier aus verantwortungsvollen Quellen
Paper from responsible sources
FSC® C105338

If you have any concerns about our products,
you can contact us on
ProductSafety@springernature.com

In case Publisher is established outside the EU,
the EU authorized representative is:
Springer Nature Customer Service Center GmbH
Europaplatz 3, 69115 Heidelberg, Germany

Printed by Libri Plureos GmbH
in Hamburg, Germany